T

Field
Guide

to

natural
phenomena

The Field Guide to

natural
phenomena

The Secret World of Optical, Atmospheric and Celestial Wonders

FIREFLY BOOKS

A FIREFLY BOOK

Published by Firefly Books Ltd. 2010

Copyright © 2010 Quid Publishing

First printing

Publisher Cataloging-in-Publication Data (U.S.)

Heidorn, Keith C.

The field guide to natural phenomena : the secret world of optical, atmospheric and celestial wonders / Keith C. Heidorn and Ian Whitelaw.

[224] p. : col. ill., photos. ; cm. Includes index.

Summary: An illustrated field guide, this book explains the science, history, natural history, and folklore behind a wide range of natural phenomena.

ISBN-13: 978-1-55407-707-6 (pbk.)

ISBN-10: 1-55407-707-9 (pbk.)

1. Science – Popular works. 2. Natural history. I. Whitelaw, Ian. II. Title.

508 dc22 QH81.H453 2010

Library and Archives Canada Cataloguing in Publication

Heidorn, K. C.

The field guide to natural phenomena : the secret world of optical, atmospheric and celestial wonders.

ISBN-13: 978-1-55407-707-6 (pbk)

ISBN-10: 1-55407-707-9 (pbk.)

1. Optics--Popular works. 2. Astronomy--Popular works.
3. Meteorology--Popular works. 4. Aquatic sciences--Popular works.
5. Electronics--Popular works. I. Whitelaw, Ian, 1953- II. Title.

Q162.H45 2010   500   C2010-902355-2

Published in the United States by
Firefly Books (U.S.) Inc.
P.O. Box 1338, Ellicott Station
Buffalo, New York 14205

Published in Canada by
Firefly Books Ltd.
66 Leek Crescent
Richmond Hill, Ontario L4B 1H1

Cover and interior design by Lindsey Johns and Luke Herriott for Quid Publishing

Printed in China

To Kristi and the Boys, and especially to Jane, Bonnie, Carol, and the rest of my Valemount "family" who supported me through my recovery year and while writing this book.

Keith Heidorn

To my mother and father, Ann and Rob, for passing on an appreciation of the beauty of language and the value of knowledge.

Ian Whitelaw

Contents

Chapter Four
ELECTRICAL PHENOMENA

Chapter Six
AQUATIC PHENOMENA

Chapter Five
GEOLOGICAL PHENOMENA

Introduction

When Louis Armstrong sang "What a Wonderful World," he acknowledged only a few of the many wonders that we can find on, above, and below Earth. From deep within the planet, where molten rock lights the fires of volcanoes and the steam of geysers, to the outermost reaches of the atmosphere where auroras play, the physical world provides us with more natural wonders than we can hope to understand in our short lifetime.

We begin the journey through these natural wonders in the incomprehensible expanse of space, looking at the billions of planets and stars and the secrets they hold—secrets that human endeavor is only just beginning to uncover. Moving closer, we consider our celestial companion the Moon and its role in some of the most spectacular events visible to us on Earth.

Our next step down brings us into the atmosphere—that layer of gas that encircles Earth, which seems so infinite and yet, when compared to the size of the planet, is only as thick as the skin on an apple. At its outer edges, the solar wind conjures up the glorious auroras. When the light from the Sun meets the atmosphere, the interaction can cause astounding wonders. As it passes through clouds of ice or water, for example, brilliant haloes, coronas, and rainbows dazzle us with their colors. When the Sun's energy heats the surface, baffling mirages can be formed, and the winds that traverse our planet are created. Some form twisting vortices called devils and waterspouts, while others pick up sand and dust. Other winds are so central to our way of life that we name them: chinook, haboob, willy-willy, and The Doctor are just a handful of these.

Toss some dust in with the water vapor in Earth's atmosphere and we beget clouds—often unnoticed, yet beautiful masses that are with us almost every day. The clouds bring rain and other precipitation including delicate ice crystals known as snow. Even when no clouds develop, water can paint the landscape with dew and frost, or sprout delicate ice flowers. At other times, clouds grow violent, flashing and rumbling unnervingly as thunderstorms rage across the sky. Even within these awesome storms, however, a natural beauty prevails in the staccato lines of lightning and the symphonic majesty of thunder.

Turning our gaze from the sky to the ground, the grandeur of Earth's varied landscapes becomes evident. Far beneath our feet, geo-fires burn, causing earth-shattering volcanoes and incredible hot-water geysers spouting like great breaching whales. The shifting of the planet's surface can produce earthquakes so powerful that they turn entire cities to rubble and leave hundreds of thousands of people dead. Other, more benign, phenomena also contribute to the changing landscape around us. The slow processes of wind and water erosion chip away at rock like a sculptor, often leaving us with wonders that rival Michelangelo in artistry. The hoodoos, mesas, buttes, spires, and arches carved around the world are monuments to the persistence of nature. Set within this natural architecture are the fossils that open a

A rural scene at sunrise is captured beautifully as the Sun's rays break through early-morning mist. Irisation (see p. 58) is visible in the clouds on the left side of the photograph.

window on life long before humans rose to preeminence. The great mountain ranges and glaciers we see today were formed over the course of millennia and bear witness to the durability of the natural world.

Finally, we come to the great waters. The atmosphere may cover all of Earth but two-thirds of what lies beneath it is water, mostly in the form of the oceans. These great waters dance beneath the pull of the moon, giving us the daily tides. Fueled by the Sun, they join with atmospheric influences to give us the global currents—siblings of the global winds. Wind and water also add the rhythm of waves to the oceans' dance, from subtle cat's paws to powerful storm surges. Occasionally the rocks surrounding the ocean basins cut in to form waves that go unnoticed as they cross the water but become devastating when they hit the land: the tsunamis.

Life and light also produce great wonders in conjunction with bodies of water, giving us eerie bioluminescence and colored oceans. Extreme and prolonged cold can cajole waters to produce further ice phenomena, from the immense glaciers to the delicacy of sea ice and the unearthly mechanism of the ice circle.

This book provides just a taste of the many wonders that occur around us. Some of these are so commonplace that we may remain oblivious to their existence. Others are rare, coming once in a blue moon—or even less frequently. Although our understanding of these are growing, there are still some that continue to puzzle scientists, such as ball lightning and the transient electrical phenomena known as sprites and jets. In the pre-scientific era, people attributed their occurrence to the work of the gods; today, the greater knowledge we boast does little to diminish the majesty of the natural phenomena as they unfold around us.

celestial

phenomena

The night sky is home to some of the most widely witnessed phenomena. From solar eclipses to "shooting stars," the ancients saw these events as either auspicious portents or harbingers of doom, and today they provide scientists with rich sources for research.

The Universe in Motion

Let us begin by contemplating the greatest natural wonder of them all, the one that encompasses all others, the one whose dimensions, whose qualities, whose very existence, stretch our minds beyond our ability to comprehend: the universe itself. Humankind has marveled at the numbers of, and our distance from, the stars since time immemorial, but in the last few decades scientific research has revealed the universe to be more remarkable than we could ever have imagined.

The sky at night is a natural wonder in its own right. If you are away from the bright lights of the city and can find true darkness on a cloudless night, gaze up and you'll soon become aware of dozens of bright stars. Most of these probably belong to the main constellations—groupings of stars that appear to form a recognizable image or pattern, such as Leo the lion, Orion the hunter, or, in the southern hemisphere, the Southern Cross (Crux) which is surrounded on three sides by the stars of the constellation Centaurus. In the Western world we owe these constellations and the signs of the zodiac to the Ancient Greeks, but other cultures have found other forms and symbols in the stars of the night sky.

Astronomers refer to these arbitrary groupings as "asterisms," and the stars within a "group" are very rarely anywhere near each other—they just look that way from where we're standing. Usually they are many light years apart, but there are exceptions. The stars that make up the Great Bear (also known as the Plow or Big Dipper) are actually relatively close together, and all three of the stars that make up Orion's belt are at similar distances from us. Each of these three stars has a mass 20 times greater than the Sun (which is, of course, a star), and they are many thousands of times brighter.

This image, composed from data accumulated by the Hubble Space Telescope over several months at the end of 2003, shows the most distant objects yet seen in the Universe. This tiny portion of the sky (the inset at the bottom left shows the viewed area in relation to the Moon) reveals some 10,000 objects, each of which is a galaxy containing many millions of stars.

THE MILKY WAY

Gaze a little longer, and you'll discover more stars, and more, and more, and if the sky is clear you'll gradually detect a denser band of stars stretching across the sky in a more or less east–west direction. These are the stars of our galaxy, the Milky Way, and there are more than 200 billion of them, bound together by gravity. The galaxy is shaped like a flat disk with two main spiral arms, and we are situated in one of the arms. Looking through the plane of the disk we see a band of light, like a trail of milk across the sky. The ancient Greeks called it *Galaxias*, from the word *gala*, meaning "milk," and we now use the word "galaxy" to refer to any gravitationally bound star system.

Because the stars are not evenly distributed throughout the galaxy, it is possible to make out lighter and darker areas within the Milky Way, and in the same way that the Greeks formed constellations from the stars, some cultures, such as the Incas, have found animal shapes in the patterns of the darker areas. The best-known "dark cloud constellation" is the Australian Aboriginal "Emu in the Sky."

THE EXPANDING UNIVERSE

Almost everything that we can see in the night sky with the naked eye—the bright stars of the constellations, the stars that make up the broad band of the Milky Way, and every one of the tiny pinpricks of light that are sprinkled across the heavens—is within our own galaxy. In the southern hemisphere it is possible to make out the Small and Large Magellanic Clouds, two galaxies on the edge of the Milky Way, and in the northern hemisphere the Andromeda and M33 spiral galaxies can just be seen, but how many others are there? The current estimate for the number of galaxies in the universe is (are you sitting down?) in excess of 125 billion! And this estimate is likely to rise as advancing technology allows astronomers to see ever fainter and more distant objects. The Hubble Space Telescope, which orbits the Earth and gazes into deep space, is continually revealing the existence of ever more galaxies in the farthest reaches of the universe.

The telescope is named after Edwin Hubble, an astronomer who made one of the most important discoveries of the last 100 years. Not only did he prove that there were galaxies beyond our own, but he also showed that all the galaxies are moving away from each other and that the greater the distance between them, the faster they are moving apart. This discovery, made in 1929, was the first empirical evidence to support the Big Bang theory, the idea that everything in the universe—time, space, energy, matter, and the laws that govern the behavior of all of these—originated at a single point and that space has been expanding ever since. It has been calculated that the Big Bang took place between 13 and 14 billion years ago, and there is evidence that the rate of expansion of the universe is accelerating.

The Milky Way is nothing but a mass of innumerable stars planted together in clusters.

Galileo Galilei

TWINKLE, TWINKLE LITTLE STAR

While you were staring into the night sky and feeling overawed and insignificant, you may have noticed that those stars actually were twinkling. This isn't because they really change their brightness. It is an effect called "astronomical scintillation," and it's caused by the atmosphere, which is one of the reasons for placing the Hubble Space Telescope in orbit outside the atmosphere.

This twinkling is the result of the light from the star passing through pockets or layers of air that have different densities, usually because they are at different temperatures. As the light passes through the boundaries between pockets of different densities—some of them just a few inches or centimeters across—it is refracted slightly,

and this changes the perceived brightness of the star. The effect is most apparent when a star is low on the horizon, as the light is passing through the greatest distance in the atmosphere.

Using infrared imagery, NASA's Spitzer Space Telescope shows the incredible density of countless stars in the central portion of the Milky Way galaxy.

15

Our Solar System

Moving a little closer to home, let's take a look at the celestial

bodies that spin around our own star—the Sun. The planets,

together with the moons, rocks, and dust that orbit them, and

the occasional flying object careening through space on

a trajectory of its own, are the elements that make up our

solar system. This "inner space" is filled with wonders,

many of which can be seen with the naked eye.

The word "planet" comes from a Greek word meaning "wander," and the planets were once thought of as wandering stars because they move in relation to the true stars. Until recently there were nine planets orbiting the Sun, but then it was decided that the smallest and most distant planet—Pluto—was too small, and it was relegated to the status of a dwarf planet, of which there are several. So now there are only eight planets. The four farthest from the Sun are composed largely of gaseous material, and they are known as Gas Giants. The four closest to the Sun, including Earth, are mainly composed of rock, and they are referred to as Terrestrials. The Terrestrials are primarily composed of a silicate mantle that surrounds a metallic core, which is composed mainly of iron

THE GAS GIANTS

The outermost planets are Neptune and Uranus. Neither of these can be seen with the naked eye, but Uranus does have the distinction of being orbited by no less than 27 moons. Next comes Saturn, the second largest planet in the solar system, and it is easily visible—if you know where to look. Of course, all the planets continuously change their positions in the sky as they orbit the Sun and our planet spins on its axis, but there are many websites, as well as astronomical software, that can tell you the position of any planet or other celestial body in the night sky at any given moment.

With the naked eye you will see just a point of light, but with even a small astronomical telescope you can make out two broad, flat concentric rings around the planet.

This image of the Sun was taken by the Extreme Ultraviolet Imaging Telescope aboard the SOHO (Solar Heliospheric Observatory) satellite. The granular appearance is caused by turbulence in the gas, which consists of double-charged helium ions at a temperature of 108,000°F (60,000°C). Plumes of particles can be seen streaming from around the Sun's edges.

The rings themselves are composed of ice and particles of rock. The gap between the rings is called the Cassini division, and it has been created by the presence of Mimas, one of at least 61 moons that orbit Saturn, whose gravity has drawn the particles to it. The biggest of Saturn's moons is Titan (which is larger than the planet Mercury), and it, too, is visible with a small telescope.

Jupiter, the innermost Gas Giant and the largest planet in the solar system, is the fifth planet from the Sun, and it is often the brightest object in the night sky other than the Moon. (Venus is actually brighter, but cannot be seen in the darkest hours—see below.) Jupiter's four largest moons—Europa, Io, Ganymede, and Callisto—are known as the Galilean moons because they were first discovered by Galileo, and they can easily be seen with a good pair of binoculars. The Great Red Spot, a giant storm taking place on the surface of the planet, can be seen with a small astronomical telescope.

THE TERRESTRIALS

Mars, the fourth planet from the Sun, was named after the Roman god of war. It can be seen with the naked eye as a steady orange-red light, but its apparent brightness varies enormously, depending on its distance from Earth. Our orbit is closer to the Sun than that of Mars, and we are traveling faster, so every 25.5 months or so we pass Mars on the inside lane, with the Sun on one side of the Earth and Mars on the opposite side. Mars is then said to be "in opposition," and it is most visible at this time, appearing in the east at sunset and staying in view all night long. The planet's opposition schedule this decade is March 2012, April 2014, May 2016, July 2018, and October 2020. Every 15 to 17 years, opposition occurs

when Mars is at the part of its elliptical orbit that brings it closest to the Sun and therefore to us. This is called "perihelic opposition," and when this occurred in 2003, Mars was as close to us as it has been in the last 60,000 years, giving us a spectacular view of the "red planet." It won't be that close again until the year 2287, so don't hold your breath, but you can very often see it clearly in the night sky nonetheless.

Named after the Roman goddess of love and beauty, Venus is the brightest of the planets, but because it is closer to the Sun than Earth is, it is never high in the night sky (meaning it is always on the sunward side of Earth). It is most commonly seen low on the horizon at dusk or, after lapping Earth, which it does every 584 days, low on the horizon at dawn. For this reason it was once thought to be two separate stars—the Evening Star and the Morning Star. However, it is so bright that it can frequently be made out during the daytime.

The orbit of Venus occasionally passes directly between Earth and the Sun, when the planet can be seen as a black dot passing across the Sun. These "transits of Venus" occur in pairs eight years apart, approximately every 130 years. The dates for the most recent transits are 1874/1882 and 2004/2012.

Last as well as least comes Mercury, the smallest of the planets and the closest to the Sun. Although it is visible with the unaided eye, it is always situated in approximately the same direction as the Sun and is therefore never seen against a dark sky, which makes it difficult to spot.

HAIRY STARS

The word "comet" comes from the Greek word for "hairy" and, like planets, comets were once thought to be stars. They appear to be hairy because they have a fuzzy atmosphere around them, known as a *coma*, and often a long, cloudlike tail. Like planets, however, they are in fact objects within our solar system in orbit around the Sun.

A comet consists of a nucleus of rock, dust, ice, and frozen gases flying through space, but this nucleus has the strange property of absorbing almost all the radiation that hits it. This means that the nucleus is very hard to see (as it does not reflect light), but it also means that the comet absorbs energy, especially when it is closest to the Sun. This in turn causes the ice and gases to melt, releasing material and electrically charged particles as the comet travels. The material forms a tail behind it, while the ionized particles are blown by the solar wind, creating a second tail that points away from the Sun.

More than 3,000 comets have been identified, but there are thought to be literally millions of them whizzing around the solar system. Almost all are too small or too distant to see, but some are spectacular and have the convenient habit of passing close to Earth on a regular basis.

Sir Isaac Newton thought it was possible that some comets might return past Earth after orbiting the Sun, but it didn't quite compute. It was his friend Edmund Halley who took into account the gravitational effects of Neptune and Jupiter and showed that a series of comets seen some 75 years apart were in fact the same comet—the one we now know as Halley's Comet. This comet, by the way, appeared in the year in which the author Samuel Langhorne Clemens, better known as Mark Twain, was born (1835) and in the year he died (1910).

Halley's Comet last passed through the inner solar system in 1986, and it will return in 2061. It is one of several that are bright enough to be seen with the naked eye, and these are known collectively as the Great Comets. Others that have been seen in recent years include Comet Hyakutake (which made a close pass in 1996 but won't be back for about 100,000 years); Comet Hale-Bopp, which had a huge nucleus some 25 miles (40km) in diameter and was seen by millions of people in North America in April of 1997; and Comet McNaught, the brightest comet in more than 40 years, which was seen in the southern hemisphere in 2007 and was visible even in daylight.

Comets have often been hailed as omens, and the appearance of Halley's Comet in 1066 was retrospectively seen as auguring the death of King Harold of England at the Battle of Hastings—the comet is depicted in the Bayeux tapestry. Even in our modern

It will be the greatest disappointment of my life if I don't go out with Halley's Comet. The Almighty has said, no doubt: "Now here are these two unaccountable freaks; they came in together, they must go out together."

Mark Twain, *Mark Twain's Autobiography*

and more enlightened times, superstition remains. When Hale-Bopp visited in 1997, it was widely believed to be accompanied by an alien spaceship and, in San Diego, 39 members of the Heaven's Gate cult ended their lives in order to board the ship and move to the Next Level before planet Earth was "recycled."

When will the next bright comet arrive? No one knows. Common comets that can be seen with the aid of a telescope are visible several times a year, but for a comet to be really bright it must have an orbit that brings it close to the Sun, so that the tail becomes obvious, and then close to the Earth so that we can see it. The orbit of such a comet is very elliptical, bringing it in from the outer edges of the solar system, so unless a comet

is "periodic," astronomers don't know it is coming until it is seen approaching. Keep an ear open for the news and take the next opportunity to see the passing of a "hairy star."

SHOOTING STARS

If you've never seen a shooting star, then you're just not looking. Given a dark and cloudless sky, the brief streaks of light made by these celestial objects can be seen at the rate of several every hour, and occasionally there are literally hundreds every hour that can be seen with the naked eye. Of course, like wandering stars and hairy stars, they aren't stars at all!

A "shooting star" is actually a meteoroid, a piece of rock or metal—ranging from a particle the size of a grain of sand to a boulder several yards or meters across, but usually about the size of a pebble—entering the atmosphere some 60 miles (100km) above Earth's surface. Some of these pieces of rock are fragments of asteroids but the majority are comet debris. Traveling at a speed of up to 150,000 miles an hour (240,000km/h), the meteoroid compresses the atmospheric gases in front of it, generating heat and causing the rock to burn up. The streak of light that this creates is called a meteor, and larger meteoroids can become fireballs that travel right across the sky and illuminate the landscape. Some meteoroids also leave a trail of ionized particles that can be seen for several minutes.

Large meteoroids sometimes explode in the atmosphere. On June 30, 1908, a powerful shock wave over the uninhabited Tunguska region of central Siberia flattened trees over an area of some 800 square miles (2,000 sq km), knocked people over more than 100 miles (160km) away, and registered on monitoring equipment in Europe and Asia. It is thought to have been caused by the explosion of an extremely large meteoroid.

At certain times of the year, meteor activity is so intense that the phenomenon is referred to as a "meteor shower." As we have already seen, comets leave trails of dust and rocks behind them, and a meteor shower occurs when Earth passes through one of these trails and literally dozens of meteoroids burn up in the atmosphere every hour. A meteor shower appears to originate from a single point in the night sky, and each shower is named after the nearest bright star or constellation to that point. The Perseid meteor shower, for example, appears to come from the direction of the constellation Perseus, but it is actually caused by debris left behind by the comet Swift-Tuttle, which returns every 130 years. The Perseid shower can be seen from the northern hemisphere for several weeks every summer, with a peak of activity in the second week of August.

The most impressive meteor shower—sometimes so intense that it is called a meteor storm—is called the Leonids, because it appears to originate in the constellation of Leo. In 1833 the storm, which was seen across much of North America as planet Earth passed through the trail left by comet Tempel-Tuttle, reached an intensity of more than 100,000 meteors an hour. The Leonids reach a peak of intensity every 33 years, but each year in mid-November a display of at least 40 meteors an hour can be seen, so keep your eyes peeled.

This photograph of Halley's Comet—taken by one of Russia's Vega probes in 1986—shows the long "hairy" tail that it leaves behind. The comet will pass close to the Earth again in 2061.

METEORITES

As they pass through the atmosphere, small meteoroids burn away completely or explode, but sometimes parts of larger meteoroids make it through the atmosphere and strike the Earth. These are then known as "meteorites," and more than 30,000 have been found so far, ranging from small pebbles to huge boulders.

The Leonid meteor shower presents its annual fireworks display around November 17. It is produced when the Earth's atmosphere collides with the debris from comet Tempel-Tuttle.

Most of these are composed of rock, but about one in 20 is made of iron or a mixture of iron and nickel. These are thought to be from the cores of disintegrated asteroids. The Hoba meteorite in Namibia is a 60-ton chunk of iron and nickel.

Several giant craters on our planet are believed to have been caused by massive meteorite impacts, and a crater off the Yucatán Peninsula in Mexico is thought to be the site of an impact that created such climatic and atmospheric disturbance that it caused the extinction of the dinosaurs some 65 million years ago. So next time you see a bright meteor, just pray it isn't another one like that.

If I had to choose a religion, the Sun as the universal giver of life would be my god.

Napoleon Bonaparte

METEORS

What they are:

Meteors—often mistakenly referred to as "shooting stars"—are streaks of light in the night sky.

What happens:

Pieces of rock known as meteoroids enter Earth's atmosphere. Traveling at high speeds, they compress the atmospheric gases in front of them, producing heat that in turn vaporizes the meteoroid, which gives off light. Meteoroids vary in size from particles of dust to large boulders, but most are about the size of a pebble. Meteor showers occur when the Earth passes through the debris from a comet tail.

Where to see them:

Meteors can be seen from anywhere on the globe on a clear night when the sky is dark (i.e., away from artificial lighting and without a bright Moon).

When to see them:

Meteors are more commonly seen between midnight and dawn than between dusk and midnight. This is because after midnight we are on a part of the globe that is facing forward on Earth's trajectory around the Sun, and any meteoroids that enter the atmosphere will do so at a higher velocity than those that are catching up with us and entering the atmosphere from the Earth's trailing edge, where the time is between dusk and midnight. The faster a meteor travels through the atmosphere, the brighter it will be. In the diagram below, we are seeing the Earth from above the North Pole and the Earth is spinning counter-clockwise. Meteoroids entering from A, in daylight, are unlikely to be visible. Those entering from B are having to catch up with the Earth, whereas the atmosphere is moving toward those entering from direction C, adding to their Earthward velocity.

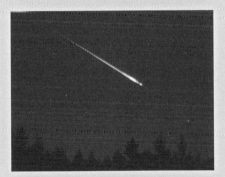

Above: A meteor streaks across the night sky as a piece of an asteroid or comet debris burns up in the atmosphere.

Below: As the spinning Earth orbits the Sun, observers in the part of the globe experiencing the pre-dawn hours are likely to see the brightest meteors.

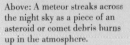

midnight

C

B

dawn

dusk

Earth's orbital path

noon

rotation of the earth

A

In so many and such important ways, then, do the planets bear witness to the Earth's mobility.

Nicolaus Copernicus, *The Revolutions of the Celestial Orbs*

THE MAGIC OF THE SEASONS

Year in, year out, the seasons come and go in an endless cycle of bloom and decay, warmth and cold, light and darkness, long days with short nights, and short days with long nights. The farther we are from the equator, the more marked are the changes that the seasons bring. This annual rhythm affects many aspects of our lives and it determines when we will experience all the different wonders that the weather provides. This may seem like a radical departure from talk of planets, comets, and meteors, but the focus of this chapter is the Sun, the objects that orbit it, and the way these things appear to us. The Sun and Earth's orbit around it are at the heart of why the seasons occur.

Every 24 hours Earth makes a little bit more than one spin, so that the same point on Earth is facing toward the Sun, and this gives us our days. And approximately every 365.25 days Earth makes a complete circuit of the Sun, which gives us the length of the year (with a leap year once every four years). If the axis around which planet Earth rotates were exactly vertical, relative to the horizontal plane of its orbit, then the midline of the planet—the equator—would always be closest to the Sun, but Earth's axis is tilted at an angle of about 23.5°, and the direction of that tilt remains constant as we orbit the Sun. As a result, in late June the northern hemisphere is maximally tilted toward the Sun while the southern hemisphere is tilted away, and in late December the opposite is the case.

The period of the year in which our home portion of the globe is tilted toward the Sun, we refer to as summer. This is the time of year when the Sun rises highest above the horizon and strikes us most directly from above, giving us longer days and imparting more heat energy. At the same time, in the opposite hemisphere, winter brings colder temperatures as the Sun sits lower in the sky and the rays are slanted across the land. The difference between summer and winter is most radical in polar regions, where the Sun does not set at all at the height of summer, and does not rise at all in the depths of winter. The seasonal variation in the intensity and duration of the Sun's energy is minimal at the equator, and the temperature variation is therefore much less. Rather than experiencing the four seasons that are commonly spoken of in higher latitudes, people in the tropics generally refer to the dry and rainy seasons, which are the most marked effect of the Earth's tilted orbit in these regions.

Between the extremes of the summer solstice (longest day) and the winter solstice (shortest day), there are the spring and fall (or vernal and autumnal) equinoxes (*equinox* means "equal night")—days when the tilt of Earth's axis is "sideways on" to the Sun, neither toward it nor away, and the Sun is directly above the equator. On these days the hours of daylight and darkness are at their most equal all over the globe.

THE SEASONS

Responding to lower temperatures and shorter days in fall, deciduous trees stop producing the green chlorophyll needed for photosynthesis, revealing the red and orange pigments in their leaves.

What they are:

The seasons are portions of the year distinguishable by such criteria as the average temperature, the length of the daylight period, the general weather pattern, and the behavior and stages of development exhibited by animals and plants.

What happens:

Due to the angle of Earth's axis of rotation, the two hemispheres are alternately tilted toward and away from the Sun in the course of a year. This results in an increase in insolation (exposure to the Sun's rays) in spring, a maximum in summer,

a decrease during the fall, and a minimum during the winter.

Where to see them:

The variation in insolation throughout the course of the year is least at the equator and greatest at the poles, but seasonal ecological changes are most obvious in mid-latitudes where there is abundant plant and animal life.

When to see them:

Maximum insolation occurs on the summer solstice (June 20/21 in the northern hemisphere, and December 21/22 in the southern), and the winter solstices are the other way around. However, these are not generally the warmest and coldest times of the year, because there is a delay in the effect of the changing intensity and duration of sunlight, as large bodies of water take time to absorb and release heat. In some latitudes the month with the warmest average temperature can be August or even September.

In its circuit around the Sun, Earth is tilted relative to the plane of its orbit. This causes variations in the amount of solar energy received across the globe in the course of a year, resulting in the seasons.

fall
spring

winter
summer

summer
winter

spring
fall

The Moon

The most obvious wonder in the night sky is, of course, the Moon. An object of mystery and romance, our only natural satellite was probably once part of our own planet. It is thought that during Earth's early history, a celestial body as big as Mars struck Earth and knocked off a huge dollop of molten rock that remained in orbit around the planet. Tests on Moon rock suggest that its composition is indeed very similar to that of the Earth.

Our silver satellite orbits Earth at a distance of some 240,000 miles (385,000km) on a slightly elliptical path, and it makes a complete lap of Earth once every 27.5 days (the word "month" is derived from the word "moon"). At the same time, Earth is spinning in the same direction as the Moon's orbit, but some 27 times faster, so the Moon passes through approximately the same position overhead about every 25 hours as we spin by beneath it.

We speak of the Moon orbiting around Earth, but in fact Earth and the Moon both orbit around their mutual center of mass, or "barycenter," which is actually 2,900 miles (4,700km) to the Moonward side of Earth's center—or about 1,000 miles (1,700km)

beneath Earth's surface on a line between the center of the Earth and the center of the Moon. The presence of the Moon therefore causes the Earth to "wobble" in its orbit around the Sun.

WAXING AND WANING

As one of the most evident wonders of the night sky, the Moon is supremely predictable. Not only do we know precisely where it will be, but we also know exactly how much of it will be illuminated. The Moon has no source of light of its own, but reflects the light of the Sun, and we can only see that part of the Moon that is facing us and is being struck by sunlight. When the Sun is on one side of the Earth and the Moon is on the other, the side of the Moon that

Low on the horizon, the full Moon in all its glory reflects sunlight across an open sea. The ancient Greeks were the first to realize that moonlight actually originates from the Sun. But the Moon is much more than a lightsource: it has a strong and intimate relationship with the world's oceans, leading to phenomena such as the daily high and low tides (see page 191).

How like a queen comes forth the lonely Moon
From the slow opening curtains of the clouds
Walking in beauty to her midnight throne!

George Croly, "Diana"

is facing Earth is fully illuminated, and this is called a full Moon; but when the Moon is on the Sunward side of Earth, we see—or rather fail to see—its unlit side. We call this a new Moon. These are called "phases" of the Moon, and phases repeat once every lunar month, which is 29.3 days. Each of these is slightly longer than the time taken by the Moon to complete one orbit of Earth, because in the course of a month Earth has completed one twelfth of its orbit of the Sun, and the Moon therefore has to travel a little farther round its orbit to be in the same position in relation to both the Sun and Earth.

The new Moon is effectively invisible from Earth, but as the angle that the Moon makes between us and the Sun changes, we start to see its illuminated portion. In the course of a month, the illuminated portion that we can see grows, or "waxes," until we see the full Moon. It then starts to shrink, or "wane," until the next new Moon occurs.

On a clear night it is possible to make out the main features of the Moon—the lighter areas and the darker patches that were once thought to be seas—with the naked eye, but a pair of binoculars or a small telescope will enable you to see much more. This Moon map shows the principal "seas" or *maria*, and the largest of the Moon's craters.

This map of the Moon shows the Latin names of some of the major features. The large dark areas were once thought to be seas (Mare Crisium, for example, means Sea of Crises), but they are now known to be huge plains of solidified basaltic lava.

THE PHASES OF THE MOON

What they are:

The phases of the moon are the changing ways the Moon appears to us throughout the course of a month. The change is a gradual continuum, but the named phases are: new Moon, waxing crescent, first quarter, waxing gibbous, full Moon, waning gibbous, last quarter, waning crescent, and back to the new Moon.

What happens:

Like Earth, half of the Moon is always illuminated by the Sun. As the Moon makes its monthly orbit of Earth, the amount of the illuminated half that we can see varies because of the changing relative positions of the Moon, Earth and Sun. When the Moon is on the same side of us as the Sun, It Is high in the sky during daylight hours and the illuminated side is away from us, so we can't see it. As the Moon moves around Earth, more of it gradually becomes illuminated until it reaches the opposite side of Earth from the Sun and we can see the whole of the lit side during the hours of darkness.

Over the course of two weeks the Moon "waxes" from the new Moon to the full Moon.

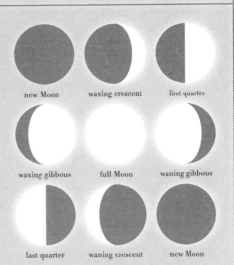

new Moon   waxing crescent   first quarter

waxing gibbous   full Moon   waning gibbous

last quarter   waning crescent   new Moon

Where to see them:

Except at the time of the new Moon, when we can see none of the Moon's illuminated surface, all phases of the Moon can be seen from any-where on Earth providing the sky is clear.

When to see them:

The Moon is always there. Most calendars and diaries will tell you the phases of the Moon.

Mother of light! how fairly dost thou go
Over those hoary crests, divinely led!
Art thou that huntress of the silver bow
Fabled of old?

Thomas Hood, "Ode to the Moon"

THE DARK SIDE OF THE MOON

Like Earth, the Moon itself is spinning on its axis, but it does so at exactly the same rate that it orbits around Earth—once every 27.5 days—and therefore we always see the same side of the Moon. The other side had never been seen by humans until it was photographed by a Soviet lunar probe in 1959. The far side of the Moon is often incorrectly referred to as the "dark side," but of course it receives as much sunlight as the side that we can see. When we are seeing the darkness of the new Moon, the far side is bathed in sunlight.

Occasionally, when there is a crescent Moon, it is possible to make out the rest of the Moon's orb very faintly. This is known as "the old Moon in the new Moon's arms." There can't be any sunlight falling on it,

so what is happening? The Italian artist and scientist Leonardo da Vinci provided the explanation back in the 16th century—Earthshine. The Moon is being lit by sunlight reflected from Earth. Given its size and the reflective qualities of clouds and oceans, Earth actually reflects much more sunlight than the Moon, and we know how bright moonlight can be.

THE MOON ILLUSION

Have you ever noticed how much bigger the Moon is when it's close to the horizon than when it's high in the sky? Well the fact is that it isn't, and you can prove this by holding a suitably sized coin at arm's length and comparing its size with that of the Moon in both instances. Whether it's low or high, the angle that the top and the bottom of the Moon make with your eye (the "angular size") is always the same. Indeed, the Moon is actually fractionally closer, and should look bigger, when it's directly above your head. So what causes this illusion?

Even though there is no sunlight falling directly on it, we can see the "old Moon in the new Moon's arms" because the Earth is reflecting sunlight onto it that is then reflected back.

Surprisingly, no one knows the answer for sure, even though the problem has been pondered over for centuries. It was thought that perhaps the atmosphere had a magnifying effect on the low Moon, but as the angular size remains the same, that's clearly not the answer. Another theory is that when the Moon is low on the horizon there are objects such as trees and buildings in the foreground that provide clues about the true distance of the Moon, and our brains therefore gain a better appreciation of its true size than when it is high in the sky where there are no clues. Or possibly the Moon appears larger when compared with the small objects on our planet than it does when seen against the vastness of the sky (as the optical illusion below, discovered by the German psychologist Hermann Ebbinghaus, demonstrates). Or it may be that when the Moon is low we actually tend to focus on the nearer objects and the Moon remains slightly out of focus, making it appear larger. The debate continues.

If you agree that the upper one of these two dark circles is larger, then clearly our perception of size is influenced by context, because they are actually the same size. Could this be the source of the Moon illusion?

THE NAMES OF THE FULL MOONS

A full Moon occurs when the Sun and Moon are on opposite sides of the Earth, and the full Moon therefore rises soon after sunset. Given the effect of the Moon illusion and the fact that the atmosphere scatters the blue light from the low Moon and gives it a reddish-orange tinge, the rising Moon can be spectacular, and it is no surprise that its arrival has been a marker of the stages in the year since time immemorial. In many cultures, the full Moon in each month has its own name, often related to the events taking place in the natural world or to the activities of the hunter or the farmer at that time of year.

The best known of the full Moons is the Harvest Moon, the one that occurs closest to the autumnal equinox—which is at the start of the fourth week of September in the northern hemisphere, and the end of the third week of March in the southern hemisphere. At this time of year, the Moon rises just half an hour later each day (during the rest of the year it is closer to 50 minutes later), giving several days in which twilight is extended by the Moon. In the days before mechanization of the harvesting process this effectively gave the farmer a chance to complete the harvest by working well into the night.

The next full Moon is commonly known as the Hunter's Moon in North America. The names of the Moon come from many sources, such as the Celtic and pagan traditions of Europe and the Native Americans, so there is a wide variety, and some shared names are applied to different full Moons. A list of common names, together with the months in which they occur in the northern hemisphere, is given on the following page.

Month	English name	Native American name
January	Old Moon	Wolf Moon
February	Wolf Moon	Snow Moon
March	Lenten Moon	Worm Moon
VERNAL EQUINOX		
April	Egg Moon	Pink Moon
May	Milk Moon	Flower Moon
June	Flower Moon	Strawberry Moon
SUMMER SOLSTICE		
July	Hay Moon	Buck Moon
August	Grain Moon	Sturgeon Moon
September	Corn Moon	Harvest Moon
AUTUMNAL EQUINOX		
October	Harvest Moon	Hunter's Moon
November	Hunter's Moon	Beaver Moon
December	Oak Moon	Cold Moon
WINTER SOLSTICE		

ONCE IN A BLUE MOON

As you can see from this list, the full Moons fall into groups of three between the solstices and equinoxes, but there is a problem. The average length of a calendar month is about 30.5 days, while the lunar month (that is, the time between successive full Moons) is only 29.3 days, so seven times in every 19 years there is an extra full Moon in the calendar year. Without a way to solve this, the Moons would gradually move back through the year—the Harvest Moon, for example, would occur earlier and earlier in the year until it finally returned to the Autumnal Equinox. Luckily, the Blue Moon comes to the rescue. In any season that has four full Moons, the third of those full Moons is called a Blue Moon, and the fourth can therefore remain what it should be. This practice has given rise to the expression "Once in a Blue Moon," meaning "on rare occasions."

And suddenly the
moon withdraws
Her sickle from
the lightening skies,
And to her sombre
cavern flies,
Wrapped in a veil of
yellow gauze.

Oscar Wilde, *La Fuite de la Lune*

Occurring around the time of the Autumnal Equinox, the Harvest Moon that lights the farmer's labor is a recurring motif in myth, literature, art, poetry, and song.

Casting Shadows

We have seen that a full Moon occurs when the Sun is on one side of the Earth and the Moon is on the other, and a new Moon is when the Sun is on the other side of the Moon—so why doesn't Earth block the sunlight from reaching the Moon, and why doesn't the new Moon stop the sunlight reaching us? The answer is that the Moon isn't usually in exactly the same plane as our orbit around the Sun—but sometimes it is.

The Moon's orbit is at an angle of about five degrees to the plane of our orbit around the Sun. This means that for most of the time the Sun, Moon, and Earth are not in the same plane and cannot possibly line up. However, twice a month the Moon does pass through the plane of our orbit, and when this occurs at the time of the full Moon—when the Sun is on one side of us and the Moon is on the other—Earth does indeed cast a shadow across the Moon. This is called a lunar eclipse.

Earth's shadow has two parts: a central cone of complete darkness, known as the *umbra*, and an outer area called the *penumbra* (meaning "almost shadow") where the sunlight is partially blocked. When the Moon passes through the penumbra, the phenomenon is called a "penumbral eclipse,"

and although these happen once a year on average, they are not very easy to detect. However, when the Moon passes through the umbra, the effect can be spectacular.

A partial eclipse occurs if only a part of the Moon is in the umbra, and we can see the edge of the Earth's shadow pass across the Moon, but in a full eclipse the Moon is completely in the Earth's shadow and is illuminated only by light that is refracted through the Earth's atmosphere. This refracted light is at the red end of the spectrum, giving the Moon a strange and eerie coloration that can range from brown through deep red, and bright red through orange, depending on the amount of cloud and dust in the Earth's atmosphere at the time. A total eclipse of the Moon can last for more than an hour and a half, and it is an unforgettable sight.

When the Moon passes between Earth and the Sun, the shadow of the Moon moves across the planet and produces a solar eclipse. Here we see an annular eclipse, which occurs when the Moon is too far from Earth to completely occlude the Sun and we can still see the Sun's outer edge.

ECLIPSE OF THE MOON

What happens:

An eclipse occurs when the Moon is on one side of Earth and the Sun is on the other, and the Moon passes through Earth's shadow. In a total eclipse, the Moon is only illuminated by light refracted around the Earth, giving the Moon a reddish appearance. In a partial eclipse, the Moon passes through Earth's penumbra, and is partly lit by the Sun.

When and where to see them:

In this decade, total eclipses of the Moon will take place on the following dates, but of course you have to be on the "right" side of the globe to see each one. In December 2010, for example, people in North America, northern Russia, and across the Arctic see the total eclipse, whereas only a partial eclipse is seen in western Europe, west Africa, and South America as the Moon sets, and in Australia and East Asia as the Moon rises. No eclipse at all is seen in India, the Middle East, Eastern Europe, or most of Africa. For precise details of where each of these eclipses can be seen, go to HM Nautical Almanac Office Eclipses Online Portal (www.eclipse.org.uk/eclbin/query_eo.cgi).

December 21, 2010	April 4, 2015
June 15, 2011	September 28, 2015
December 10, 2011	January 31, 2018
April 15, 2014	July 27, 2018
October 8, 2014	January 21, 2019

Above: Illuminated only by light that has been refracted around Earth, the eclipsed Moon takes on a strange red or orange coloration.

Below: Lunar eclipse geometry

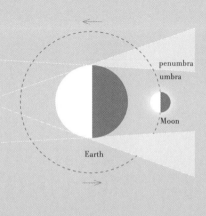

penumbra

umbra

Moon

Earth

Sun

A TOTAL ECLIPSE OF THE SUN

Now imagine the Moon crossing the plane of our orbit around the Sun at the time of the new Moon. The Moon is now directly between the Sun and Earth, and it does cast a shadow, but the situation is very different from a lunar eclipse and the result is even more spectacular. Whereas Earth's shadow is large enough to more than cover the Moon, the shadow of the Moon on the Earth is only about 100 miles (160km) in diameter. As the three celestial bodies change their relative positions, this shadow moves across a portion of the globe, and anyone standing on this "Path of Totality" will see a total eclipse of the Sun. (In the zones on either side of this path, observers experience a partial eclipse, in which the Moon occludes part of the Sun.)

Seen from any particular point on Earth, a total eclipse lasts only a few minutes, and the phenomenon is dramatic, especially in the middle of a bright and cloudless day. As the Moon moves across the Sun, the light begins to fade, with the sky darkening from the west until the whole sky is dark blue, Earth is in twilight, and the black disk of the Moon has completely replaced the fiery ball of the Sun. Now, in the brief period of the total eclipse (the only part of an eclipse that can safely be looked at without the aid of special filters to protect the eyes), something remarkable happens.

By complete coincidence, when it is at the nearest point to Earth on its elliptical orbit, the Moon is just the right size to block out the Sun completely—no more and no less—and this provides a unique opportunity to see what is happening around the Sun without being blinded by the light. At the peak of the eclipse, glowing white wisps become visible extending out from the edges of the Moon's disk. This is the Sun's corona, a pulsing, moving atmosphere of ionized gas some 200 times hotter than the surface of the Sun itself, throwing out loops and plumes of plasma that can be seen with the naked eye—but only during a total eclipse.

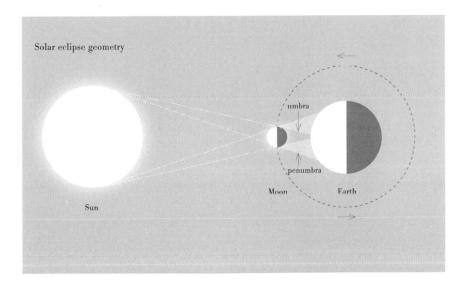

Solar eclipse geometry

umbra

penumbra

Moon   Earth

Sun

Occasionally, solar flares and prominences—huge fiery explosions erupting from the Sun's surface—can also be seen.

OTHER FORMS OF SOLAR ECLIPSE

The Moon's orbit around Earth is elliptical, meaning that its distance from Earth changes as it passes along its orbital path. So what happens when an eclipse occurs with the Moon at its farthest point from Earth? The answer is that it is not large enough to completely cover the Sun, and along what would have been the Path of Totality observers see the outer edge of the Sun forming a ring around the Moon. This is called an "annular eclipse," and no part of this eclipse can be viewed without adequate eye protection. Occasionally, the Moon causes a total eclipse at some places on Earth and an annular eclipse at others. This is known as a "hybrid eclipse."

When the Moon's shadow passes close to the Earth but not across it, some parts of the world will see a partial eclipse but no total eclipse can be seen.

Seen from Earth, the Moon is a similar size to the Sun, but seen from farther away (in this case from an Earth-orbiting satellite) it appears as a smaller black disk as it passes across the Sun.

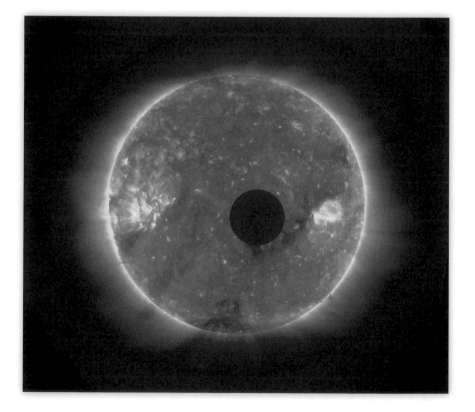

ECLIPSE OF THE SUN

What happens:

During an eclipse of the Sun, the Moon passes
between the Sun and Earth, and for observers
along a narrow, 100-mile (160-km) strip, known as
the Path of Totality, the Moon occludes the Sun,
bringing a spooky twilight and making it possible
to see the corona that surrounds the Sun. When
the Moon is at its farthest point away from the
Earth, it does not block out the whole of the Sun,
and an annular eclipse, in which the outer edge of
the Sun remains visible, takes place.

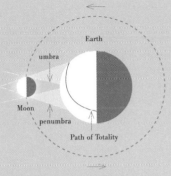

When and where to see one:

Whereas a lunar eclipse can be seen over a
large proportion on the globe, a total or annular solar eclipse occurs only over a relatively
narrow path. The Path of Totality of the total
eclipse in 2012, for example, passes mainly over
the southern Pacific and only clips the northern
parts of Australia's Northern Territory and
Queensland. The schedule of eclipses for this
decade is as follows:

Date	Type	Path of Totality
July 11, 2010	Total	South Pacific, southern tip of South America
May 20, 2012	Annular	China, Japan, North Pacific, western U.S.
November 13, 2012	Total	Northern Australia, South Pacific
May 10, 2013	Annular	Northern Australia, Central Pacific
November 3, 2013	Hybrid	Atlantic, central Africa
April 29, 2014	Annular	Antarctica
March 20, 2015	Total	North Atlantic between Iceland and Scandinavia
March 9, 2016	Total	Indonesia, North Pacific
September 1, 2016	Annular	Atlantic, Central Africa, Madagascar, Indian Ocean
February 26, 2017	Annular	Pacific, Chile, Argentina, South Atlantic, Africa
August 21, 2017	Total	North Pacific, U.S., North Atlantic
July 2, 2019	Total	South Pacific, Chile, Argentina
December 26, 2019	Annular	Saudi Arabia, India, Sumatra, Borneo
June 21, 2020	Annular	Central Africa, South Asia, China, Pacific

optical

phenomena

Between the heavens and the Earth can be found a myriad of glorious optical wonders at play in the atmosphere. Some, such as red sunsets, form in clear skies. Others, such as rainbows and haloes, occur when sunlight passes through raindrops or ice crystals.

Light from the Sun

When humans first lifted their eyes skyward, they assumed that the heavens above spread like a ceiling across the Earth. They did not understand the great distances between the heavenly bodies and Earth. The Moon and planets, the Sun and stars, they believed, sat along an outer sphere that rotated around the Earth once a day, not much farther away than the clouds. As a result, they believed the optical wonders of the sky we know today resided in the heavens, and often gave these wonders spiritual significance.

Today, we know that all atmospheric optical phenomena take place within the relatively short distance of just a few thousand miles, rather than the astronomical distances of those heavenly bodies above them; most occur within 15 miles (24km) of the Earth's surface. Only auroras take place at the tenuous edge of the atmosphere.

In this chapter, we will look at many different natural wonders that arise when the light from the Sun interacts with atmospheric gases and particles. Initially, our tour will look at those wonders arising from light's interaction with very small atmospheric constituents: the gases and small dust particles that give color to the clear sky. Included here will be a look at the world of mirages that arise from differences in the density of the air through which the light travels.

Next, we consider light's interactions with very small water droplets, such as those found in clouds, and we'll follow that with a look at the effects of the larger drops that form rainbows. We then move from liquid to frozen water to see how light interacts with ice crystals floating in the air. Finally, we jump to the top of the atmosphere to consider the fascinating wonder of the auroras. First, however, we need to look at sunlight and a few principles of optics that are common to all optical phenomena.

The Sun shines from behind a small cumulus cloud, creating a silver lining against the darker interior. Breaking through a gap in cloud, crepuscular rays are also visible. The energy produced by the Sun is most intense in the wavelengths that are visible to human eyes, which, together, appear white despite being a mixture of all the colors in the spectrum.

Know the signs of the sky and you will far the happier be.

Benjamin Franklin

A MIXTURE OF WAVELENGTHS

All natural optical wonders begin with light from our Sun. (Even moonlight is reflected sunlight.) This provides the color pallet found in all these wonders, but rather than sending us nice individual packets of color, the Sun sends its light in a mix of radiant energy, predominantly in the short wavelengths of the electromagnetic spectrum. The Sun's radiant energy has its peak intensity in the "visible light" wavelengths, so called because they are visible to human eyes. Solar light appears white, but it is actually a mix of all color wavelengths from violet to red.

When sunlight reaches the atmosphere, some is reflected by the atmosphere's atoms, molecules, and particles, and by clouds floating in it, and is sent back into outer space. A fraction is absorbed by atmospheric constituents and is used to alter their physical properties. The remainder continues down to the surface, with some being scattered off the primary path by dust and molecules. The

planet's surface can reflect the remaining radiant energy, both the direct beam and the scattered rays, back through the atmosphere, or the energy can be absorbed by the surface and turned into heat.

So as sunlight traverses the atmosphere, there is a range of optical processes that may alter its path and nature. The first process, called "scattering," diverts some light rays from their path onto an altered path. The second, refraction, bends light off its direct path as it passes from one medium to another of different density. When refraction is great enough, the white light beam will be visibly split into its constituent colors. Another form of "light bending" is diffraction, caused by light waves passing around atmospheric constituents of a similar size. Finally, sunlight may be reflected, "bouncing" off a surface such as an ice crystal or raindrop at an angle equal to that at which it struck.

Now let's look at the natural wonders these processes produce.

Long grass is silhouetted against a fiery red sunset that ignites the underbelly of stratocumulus clouds. It is thought that pollution in the atmosphere contributes to the intensity of sky colors at sunset.

Clear-Air Phenomena

Many powerful natural wonders can be found right before our
eyes, even in the clear air around us. They have inspired poets
and painters, storytellers and songwriters, and yet we often
overlook these wonders unfolding before us because we fail
to see them. Clear-air wonders include the daily parade of sky
colors from dawn to dusk, the elusive, fleeting green flash,
piercing crepuscular rays, and baffling mirages.

We begin the discussion of atmospheric
optical wonders with their simplest
forms: wonders of the clear air. By
"clear air" we refer to air that contains only
gases and particles smaller than ice crystals
or water droplets (whose influences we shall
address in later chapters). These clear-air
optical wonders require only sunlight, our
atmosphere of mixed gases, and varying
amounts of dust or similar minute particles to
manifest themselves. In the case of mirages,
they only need the atmosphere to layer itself
into segments of different densities.

Although the constituents of clear-air
optical wonders are simple, the processes
involved in their formation can be quite
complex, and some, such as the green flash,
are not yet fully understood. Indeed, the science behind most has only been explained
in detail in recent centuries. Since pre-
history, however, clear-air wonders have
inspired the arts and laid the foundations for
myths and beliefs that have evoked awe and
fear. Blood-red skies have instilled terror,
and golden sunsets have inspired spiritual
awakenings. Sky color has also been used
as the basis for early weather proverbs and
sayings—the precursors of weather forecasts.

Crepuscular rays have been interpreted as
stairways to heaven, while mirages have
fooled many a desert traveler or seafarer.

WHY THE SKY IS BLUE

At first glance, the oft-asked question "Why
is the sky blue?" seems simple. However,
images sent back from Mars and the Moon
show us that the sky can also be red or
black. So perhaps the question should be
rephrased to ask, "Why is *our* sky blue?"
The answer lies in our atmosphere.

When the Sun's light reaches our atmosphere, it encounters a mixture of atoms and
molecules with assorted dust particles mixed
into its lower regions. During the split second
it takes the solar rays to reach our eyes at the
surface, these atmospheric constituents can
alter the path of some incoming solar rays.
Many are bumped off the direct path and
scattered, continuing their downward journey on an angle different from the original
direct solar beam.

We find the answer to sky color within
this scattered portion of the solar beam.
Scattering affects each wavelength of the
solar beam, which contains all the spectral
colors, differently. The final effect depends

A roseate redness, clear as amber, suffuses the low western sky about the sun . . . The atmosphere there is like some kind of wine, perchance, or molten cinnamon, if that is red, in which also all kinds of pearls and precious stones are melted.

Henry David Thoreau, *Journals*

on the size of the particle doing the scattering. Air atoms and molecules, being relatively small compared to other particles in the atmosphere, scatter the shorter solar wavelengths (the violets and blues) more than the longer visible wavelengths (the yellows and reds).

While some light rays from the solar beam (produced by the sun) scatter in various directions, the remaining direct beam continues downward on its original course. The scattered rays are what we call "skylight." They generally contain more blue wavelengths than the direct solar beam, which usually appears whitish—a combination of all visible wavelengths. Therefore, much of the light coming to us from skylight is a shade of blue, becoming whiter the closer to the Sun you look. The sky has darker blue hues away from the direct solar beam and, near the horizon, it appears more milky-gray as other colors contribute to the mix.

A brightly colored sunset is captured from Sutro Bath at Land's End in San Francisco. Note the sky's colors reflected in the water.

In stark contrast, with no atmosphere and thus no scattered light, the Moon's sky is black. Similarly, the higher we ascend from the Earth's surface, by either climbing a mountain or flying in an aircraft, the darker the sky appears away from the solar beam, eventually becoming the black of outer space. This occurs because at higher altitudes there is less atmosphere to scatter the light than there is at the surface.

The Sun's elevation angle as it traverses the heavens from dawn to dusk is also a factor in determining the impact of scattering. When the Sun is directly overhead (local noon), the amount of atmosphere the solar beam must traverse is at a relative minimum. But when the Sun is low on the horizon and the light is traveling across the surface of the Earth, the relative "optical thickness" of the atmosphere is dramatically increased—about 38 times greater than when it is directly overhead. As we shall see, this affects the color of the sky at dusk or dawn when much of the blue light is scattered away. In fact, we can state a general rule for the change in color with increasing optical thickness: The more air a light beam must penetrate, the redder (and less blue) it becomes.

The above discussion assumes that the atmosphere consists mainly of small gas molecules such as nitrogen and oxygen. Water vapor, other larger gas molecules, and dust make up only a small fraction of the total atmosphere, but they, too, can have a significant influence on sky color, particularly when concentrated in one portion of the atmosphere between the Sun and the viewer. Water-vapor molecules, although smaller than nitrogen or oxygen molecules, have a great tendency to cling to or combine with minute solid particles or other gases, thereby forming larger particles that scatter the longer-wavelength colors—the yellows and reds—as well as the shorter wavelengths. As a result, they produce alternative hues to the dominant blues, ranging from murky grays and browns, common in polluted urban atmospheres or downwind of wildfires, to vivid oranges and reds seen at low Sun angles during sunrise and sunset.

Particles of volcanic ash and gases sent high into the atmosphere by violent explosions can remain suspended in the atmosphere for many years and push the sky's blue color to a more reddish hue such as pale red-violet or mauve, particularly around twilight. An atmosphere filled with such dust can appear deeply red. Myths and folk beliefs fearing blood-red skies probably grew out of the effects of volcanic dust on sky color.

TWILIGHT COLORS

"Red sky at night, sailor's delight; Red sky in morning, sailors take warning." This piece of weather lore and similar ones can be found in writings as far back as the ancient Greeks, including in works by Theophrastus (*On Weather Signs*), the teachings of Jesus (Matthew 16:2–3), and the poems and plays of Shakespeare (*Venus and Adonis*). According to the artist, a blood-red sky near sunset inspired Edvard Munch to paint *The Scream*.

Indeed, this saying has some scientific truth behind it as a short-term forecaster, particularly the first phrase. In both hemispheres, mid-latitude weather systems usually come from a westerly direction, so a red sky to the west (in other words in the direction of the sunset) is evidence that there is no stormy weather immediately downwind.

Also, a very red sky in the evening, rather than a pale red or even yellow sunset sky, indicates a high concentration of dust particles in the air. High-pressure cells, which contain higher concentrations of dust, do not usually produce stormy weather. On the other side of the saying, a red sky in the morning is indicative of a high-pressure cell extending to the east, and since stormier weather generally follows the high-pressure cell, one must take warning that a storm might be approaching.

During the twilight period of the day, when the Sun is on the horizon, the clear sky surrounding the solar disk takes on an orange-yellow glow. The colors in the red/yellow end of the spectrum dominate because the optically thick intervening air has scattered out all the blue wavelengths from the sunlight before it reaches our eyes. Dust and other particles enhance the reddening, and the more large particles in the air mass between the Sun and the observer, the redder the sunset/sunrise. The effect can be more dramatic when clouds are also present in the twilight sky, as they add texture to the sky color, producing varying shades, light patches, and dark shadows.

As the Sun sets, the clear sky above it glows a pale yellow with a yellow-orange aureole surrounding the solar disk. A blue-white arch tops the aureole—the twilight arch. As twilight progresses toward sunset, the twilight arch becomes pink with yellow and orange beneath. When the Sun drops below the horizon, only the red wavelengths remain, often producing a coppery or blood-red twilight arch. As twilight continues, the twilight arch slowly flattens, and the sky above darkens from blue-gray to deep blue before merging into the darkness of night.

If we turn our backs to the setting Sun, we see other effects in progress: the rising Earth shadow, the darkest region in the eastern sky, and the anti-twilight arch. The setting Sun casts the Earth's shadow onto the atmosphere, appearing as a large blue-gray sector, perhaps tinged with violet. The Earth shadow appears to rise rapidly, and its edge disappears when the shadow limb reaches about 10–15 degrees above the eastern horizon. The anti-twilight arch, which divides the Earth's shadow from that part of the sky still lit by direct sunlight, has a faint purple or yellow tint. It starts as a thin line stretching across the sky opposite the Sun and rises as the Sun sets. As the Sun descends below the horizon, the anti-twilight arch rises, becoming less distinct until it finally blends smoothly into the dark night sky.

Here, the sequence of events for the sunset twilight period has been detailed, but remember that a similar sequence takes place in reverse (in time and sky position) during the pre-sunrise twilight.

Sundown is the hour for many strange effects in light and shade—enough to make the colorist go delirious—long spokes of molten silver sent horizontally through the trees . . .

Walt Whitman, *Leaves of Grass*

THE GREEN FLASH

At the beginning or end of the day, as sunlight's first or final glimmers reach the horizon, the uppermost limb (edge) of the solar orb may appear to be tinted greenish-yellow. On rare occasions, and usually when sunset or sunrise is viewed over water and the air is clear, a momentary "green flash" may be seen on the top of the solar disk just before it emerges or disappears from view.

For many years, the green flash floated in the realm of legend and optical illusion (a trick of the mind), but we now know that this fleeting phenomenon is real, as it has been photographed by many observers. Questions remain as to exactly how it forms, although there are strong theories.

The green flash phenomenon can be seen when a small part of the Sun's orb suddenly changes color from red, or orange, to green during the orb's final second before dipping below the horizon (or in opposite sequence as the Sun emerges above the horizon at dawn). The word "flash" is

A fleeting green flash tops the setting Sun. Its solar disk has been distorted to an oval shape by greater refraction at the horizon as its light moves from outer space into Earth's atmosphere.

appropriate as the phenomenon appears suddenly and usually lasts no more than a second. There appear to be two distinct forms of the green flash. The most common manifests as a green "dot" that tops the setting or rising Sun just as it hits the horizon. The second, rarer form is more truly a "flash," emanating as a brief ray or glow of green, appearing to shoot upward from the sunrise/sunset point.

In the most widely accepted basic explanation, refraction—the bending of light as it moves from one medium into another of different density—causes the green flash. In this case, one medium is the near-vacuum of outer space and the other, denser medium

is our atmosphere. The atmosphere acts like a prism and not only bends the sunbeam from its original path, but bends each color in the white-light beam a different amount, red light being refracted the most and blue light the least. The more atmosphere the light must penetrate, the greater the effect, so the refraction of light is greater when the Sun is on the horizon than when it is overhead.

Assume for a moment that a sunbeam is composed of only red, green, and blue light. With extreme refraction, we could theoretically see three distinct suns: a red orb with a green one above it, and a blue one higher still, rather than the single true orb. As the level of refraction decreases, the three distinct orbs begin to overlap. Just prior to completely merging into a single orb, a small limb of green orb would be visible above the red orb and a small limb of blue above them both. The blue, however, would be diminished by scattering and be unseen unless the air was very clear. What we would see as the red orb sank below the horizon, therefore, is a brief glimpse of the green orb's top limb. In reality, with full-spectrum sunlight and the degree of refraction found at sunset, the actual sequence is a continuous shift in the color of the upper solar orb from red through to blue with only the briefest display of green, producing a green flash.

CREPUSCULAR RAYS

Most of us in our youth drew pictures of the Sun shining over the landscape with rays emanating from the glowing orb. We probably called them "sunbeams," but other cultures have given them more stirring names, such as the Rays of Buddha or Jacob's Ladder. To the scientist they go by the name of "crepuscular rays," meaning "rays related to twilight," although they can be seen anytime the Sun is in the sky. Appearing as either light or dark shafts that radiate out from the Sun, crepuscular rays are most commonly seen, and often at their most beautiful, around sunrise and sunset.

To see crepuscular rays we need only three elements: the Sun, something to cast a shadow, and a little dust or other particles in the air to make the rays visible. In the high sky, clouds are the usual shadow makers, although mountain ranges may also paint sunbeams across the sky when the Sun drops behind them. The critical role of the shadow-casters is to break up the sunlit sky into regions of light and dark. However, solar light rays are not visible on their own, and they require the assistance of small particles—dust, water droplets, even snow or ice crystals—to scatter or reflect the light toward our eyes. We have all seen dust dancing in a sunbeam as it shines through a window; that dust provides the effect we seek to make crepuscular rays visible.

Just a touch of scarlet

Lighting up the rosemary

The early dawn paints the clouds left by the summer rain

A color crayon morning slips softly through the window glass.

Gary Smith, "Color Crayon Morning" (song lyrics and poem), *Windsinger*, 1976

Generally, crepuscular rays appear at their brightest when a dark shadow, such as a dark cloud, provides the background, giving the greatest contrast to the rays' brilliance. A dark background may also bring out vivid colors in crepuscular rays produced when red-yellow wavelengths dominate the sunlight. Crepuscular ray strength also depends on the density of the scattering particles, the angle of the Sun, and the line-of-sight distance through the light ray. The highest contrast occurs when we view crepuscular rays looking toward or away from the Sun.

Crepuscular rays appear to converge on the Sun, but in fact they are parallel. This is due to perspective—the apparent convergence of parallel lines at some distant vanishing point, the same visual effect that causes railroad tracks or telephone lines along both sides of a highway to appear to converge in the far distance. When we look toward the Sun, we see the crepuscular rays diverging away from the Sun. However, if we turn around, we may see them converging again in the eastern sky. These rays, which appear to converge on the anti-solar (opposite the Sun) point of the sky, are called anti-crepuscular rays. When visible, "anti-crepuscular rays" generally appear as a pastel-pink glow against the darker blue sky.

Crepuscular rays stretch out from behind cumulus clouds in Plano, Texas. Though such rays appear to converge on the Sun, they are actually parallel.

. . . molding the bluffs and houses of the opposite shore into wondrous castles that, alike, tower into the sky and sink beneath the surface; nor is it strange that this poetic name should become generic, as it has, for all such multiple mirages, whenever they occur.

William J. Humphreys on Fata Morgana, *Physics of the Air*, 1920

MIRAGES: CHALLENGING THE IMAGINATION

For many the word "mirage" brings thoughts of thirsty travelers trudging slowly across desert sands toward the image of a pool of "cool, clear water," to quote a phrase from a popular country and western song. Perhaps our image is one of cowboys in the U.S. southwestern desert or of French Legionnaires lost in the Sahara Desert. Or perhaps "mirage" might conjure up a vision of fairy castles or ghost ships floating over the sea. Mirages are often associated with figments of the imagination, illusions, or hallucinations that are consigned to story and legend along with fairies and magicians. Mirages are, in fact, real-world images that can be photographed. They are refractive phenomena in the atmosphere that displace the image of some distant object from its true position. The resulting image may also be distorted or inverted, or may waver in and out of sight.

For a mirage to form, the atmosphere must have a large density gradient through a portion of it or a layer of differing density to refract the light. Strictly speaking, a density gradient is a layering of air of different densities atop one another, but most often density gradients can be simplified to temperature gradients—the change of air temperature with altitude—as a surrogate measure. Colder air is denser than warmer air: the greater the temperature contrast, the greater the density difference.

Mirages form when light rays emitted from a source or reflected off an object are bent—refracted—as the path of the light ray crosses the interfaces between air layers of different densities. Since cooler air is denser than warmer air, the light rays bend toward the cooler air, that is, with colder air on the inside of the bend curvature.

We can partition atmospheric mirages into two types: inferior and superior. (Another type of common mirage, the astronomical mirage, results from refraction of extraterrestrial light as it enters the atmosphere, as in the discussion of the green flash.) The terms "inferior" and "superior" do not comment on the quality of the mirage, but on the relative position of the mirage image in relation to the actual object's true position. Inferior mirages appear below the object's true location. Superior mirages are seen above the true location. This shift in location is further enhanced by our brain's assumption that a light ray must move in a straight line from source to eye, and this imparts the mystical, or illusional, quality to a mirage.

INFERIOR MIRAGES

The inferior mirage is the classic "desert" mirage, though perhaps today we should call it the "highway mirage," that pool of water we see ahead of us on the road but never reach.

Inferior mirages form when light rays pass through a relatively hot layer of air near the ground, usually produced by strong solar heating. In this situation, the temperature gradient has much warmer air below and cooler air above. The hotter the surface air is relative to that above, the greater the bending effect. Above a paved road, the full Sun can generate surface temperatures 20–30°F (11–17°C) hotter than the air only inches above. The common inferior mirage we see on the highway is a classic case of our eyes seeing something and the mind incorrectly interpreting the sight. The light rays from the sky are bent back upward near the surface, which is much hotter than the air above it, toward the viewer's eyes. However, the brain "sees" the ray's path as a straight line and assumes the object lies

on the surface when actually we are seeing a piece of the sky. Its bluish, often rippled appearance, however, is interpreted as being a body of water.

An inferior mirage image can also be distorted in a number of ways. The image may appear taller than normal or stretched, a condition known as "towering." Or it may appear shorter or compressed, a condition called "stooping." Complex refractions through multiple changes in density between the object and viewer may cause both towering and stooping of different portions of a large object. The formation of multiple images is also possible, creating further unusual scenes. When such complex conditions occur, you can often see

An inferior mirage gives the viewer the illusion that a large pool of water covers the surface ahead. Inferior mirages are so-named because the image is below the true object.

different views of the object by altering your eye level, moving it either up or down to catch different light-ray paths.

SUPERIOR MIRAGES

The superior mirage occurs under the reverse atmospheric conditions from the inferior mirage. In the basic case, the air close to the surface must be much colder than the air above it. This condition is common over snow, ice, and bodies of cold water. Moving through such a temperature gradient, the light rays bend downward toward the surface. Our minds interpret the vision as the object floating in the air—for example, a boat appearing to sail in the clouds. The superior mirage may cause the image of an object, or parts of it, to appear:

- visible even though the object is actually located below the geometric horizon, a phenomenon called "looming"

- lifted well above its actual position, as if floating in the air

- inverted from its normal image

- multiplied and either upright or inverted

- taller, larger, or closer than it actually is (towering)

- shorter, smaller, or farther away than it actually is (stooping).

The most famous superior mirage is the *Fata Morgana*, named after Morgan Le Fay, King Arthur's enchanted half-sister who lived in a crystal castle beneath the sea which, on occasion, appeared to rise above the waves. Italian poets named this vision in the early 19th century when they saw it over the Strait of Messina. In reality, the Fata Morgana is a multilayered superior mirage

in which erect and inverted images appear above one another. These mirage conditions can transform distant objects, even objects as relatively smooth as water or snow surfaces, into spikes or towers rising up with great vertical exaggeration.

Some historians believe that another form of superior mirage, the Arctic mirage, may have played a role in the early exploration of the Arctic, perhaps even helping the Vikings reach the shores of Greenland and North America. Under the extreme refraction of an Arctic mirage condition, topographical features—such as mountains and glaciers—lying below the horizon may rise into view, overcoming restrictions on distant viewing imposed by Earth's curvature. While a 3,000-foot (1,000-m) mountain normally disappears about 72 miles (125km) away from an observer, it may remain visible beyond 250 miles (430km) under Arctic mirage conditions. For example, under mirage conditions, Iceland's 6,600-foot (2,000-m) Vatna glacier becomes visible from the coasts of the Faeroe Islands 240 miles (385km) distant.

The Arctic mirage may have influenced the mythologies and legends of the polar region's native communities, as well as the early European view of the northern regions. Under specific, but quite common, weather conditions conducive to the Arctic mirage, the bending of light rays can give the impression that the Earth's surface is flat or turned up rather than having its true rounded downward slope.

SUPERIOR MIRAGE

What they are:

Superior mirages are mirages for which a perceived image forms above the actual location of the original object. They can make objects appear to be floating in the air and cause objects actually located below the horizon to appear above it, a condition called looming. The superior mirage can also cause objects to appear inverted. Images may appear taller than they actually are, called towering, or shorter, called stooping. In certain situations, multiple images may be seen.

What happens:

Superior mirage occurs when the air layer close to the surface is much colder (denser) than the air above it, a condition common over snow, ice, and cold waters. Light rays traveling through these different air layers bend downward toward the cold air, thus making an object appear higher or taller than it actually is.

This exaggerated schematic diagram indicates how a surface object may appear elevated. The solid line indicates the approximate path of light from the boat. The dashed lines indicate an interpretation of the image, assuming that the light rays are straight.

Where to see them:

Most frequently observed over water surfaces colder than the air above, though they can be seen whenever the air near the ground is colder than the air above, such as over a snow cover. Theoretically, therefore, they can be seen anywhere if atmospheric conditions are right.

When to see them:

More frequent in winter, at night, or in the morning when cold air hugs the surface, but can be common in the summer and afternoon over colder waters.

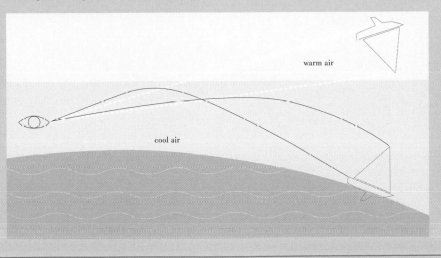

warm air

cool air

Water-Droplet Phenomena

When light passes through small water droplets, such as those found in clouds and fog, optical wonders may emerge that, in the past, were ascribed spiritual significance. One of these, the "glory," draws its name from the holy halo or corona found in many religious paintings. Yet the inspiration for such religious imagery may have originated from the bright rings often visible around the Sun and Moon, which are called the *corona.*

The leading cause of an unclear sky is the collection of billions of small water droplets that produce visible clouds. Cloud droplets generally have diameters in the range of between 0.01 and 0.02 millimeters, with the largest reaching 0.15 millimeters. Typically, it takes a million cloud droplets to form the smallest raindrop—a drizzle droplet—with a diameter of 1.2 millimeters.

Such small droplets have a size comparable to, or slightly larger than, the wavelengths of visible light and they can therefore interact with the light waves through a process known as "diffraction." Diffraction occurs when a light wave's path bends around an object or objects of a simi-

lar size to its wavelength. As a wave, light can interact with other light waves, forming combination patterns. The most important interactions are constructive and destructive interference. In constructive interference, the two waves add to produce an enhanced condition characterized by increased brightness of bands or spots. In contrast, destructive interference degrades the overall light quality, resulting in dark bands or spots.

The interaction of light with small water droplets through diffraction produces two distinct types of natural wonders, which characterize coronas and glories. Corona-type phenomena result from light shining through the particle field as it moves from source to viewer. Glory types appear when

Irisation, or iridescence, in clouds and a corona around the sun (visible at the center) produce a striking display of sky colors. Irisation manifests as brilliant pastel colors on clouds around the sun or some distance from it. These phenomena are caused by the diffraction of light around small particles such as cloud droplets.

light is reflected or scattered back through a volume of droplets to the viewer.

THE CORONA

Two optical wonders that occasionally encircle the Sun or Moon are the corona and the halo. While they are often confused, they are quite different in appearance and of very different origins. The corona (Latin for "crown") hugs rather tight against its light source and results from light interactions with small water droplets or large particles (diffraction), while haloes spread wide and arise from light rays passing through ice crystals (refraction).

The ideal corona tightly encircles the Sun or Moon with a bright and nearly white inner aureole—the combining of all color wavelengths—fringed with rings of color. When viewing conditions are optimal, the outer corona is composed of several subtly colored rings encircling the central aureole. These gradations begin with a bluish innermost ring then shift through the spectrum with green and yellow rings to the outer red ring. The most commonly seen coronas, however, have a light bluish aureole with a reddish ring surrounding it.

Coronas arise from the interaction of sunlight or moonlight with cloud water droplets. They can also form in the presence of small ice crystals and other airborne particles such as dust or pollens, which need not be transparent or even spherical in shape. The reason that non-transparent particle types can form coronas is that the small particles diffract the light waves as they pass around and not through them.

The particle sizes required to form coronas must fall within the diameter range of 0.02 to 0.1 millimeter to properly interact with the wavelengths of visible light.

The most distinct coronas occur when particles are of uniform size. The more diversity there is in the particle sizes, the more blurred the composite diffraction effect. The smallest particles generate the largest coronas and usually the brightest. Larger particles produce smaller diffraction angles and thus tighter coronas. Altocumulus clouds are most commonly associated with coronas because they generally have a more uniform droplet distribution during their rather short lifetime.

IRISATION

An extremely beautiful cousin of the corona is called "irisation" or iridescence. It often resembles an oil slick. Irisation need not ring the light source but can form brightly colored patches at larger angular distances from the Sun or Moon. The patches may be metallic- or pastel-hued, predominantly in the pink, purple, and light green shades, although shades of deeper blue and red may also appear. The variety of hues derives from the mixing of several diffraction bands caused by very small (in the micrometer range) droplets of a uniform size.

The most common cloud types associated with bands of irisation are altocumulus lenticularis, cirrocumulus (see pp. 96–7), and occasionally fractocumulus. Irisation bands are usually found at some distance from the Sun or Moon, as much as 45 degrees off the light source, which can lead them to be confused with some halo phenomena.

OF GLORY AND THE BROCKEN SPECTER

Several monks stroll along a narrow, cloud-enshrouded, high-mountain trail in quietude—each has taken a vow of silence. The lead monk stops abruptly and peers toward the mountains ahead. There, on the cloudbank before him, he sees his shadow encircled by a halo! His spirit soars! Has he found enlightenment? Has God chosen him for glory? Looking back on his fellow monks, he sees that each has a smile of serenity, for each is seeing a halo around his own shadow but not around those of his brethren.

The monks' experience was not a common one in those distant times, for only the religiously devout lived in such high realms and only the most adventurous traveled to such heights. Such visions eventually became known as the "glory" because the encircling haloes resembled the haloes given to signify the glorification of the chosen in religious art.

But glories are not heavenly signs; they are natural wonders. Today, we can view glories from aircraft flying in sunlight above uniform cloud decks. With the sun high above the aircraft, the glory forms around the aircraft's shadow viewed on the cloud deck thousands of feet below and racing with the aircraft across the sky. Although a glory results from similar physical processes to the corona, it forms from light scattered back toward the observer through the droplets rather than arising from light from an object passing through the droplets before reaching the observer.

The glory becomes visible only when sunlight is at the observer's back; therefore, it is always exactly opposite the Sun, centered on the anti-solar point. A glory appears if that light enters into a cloud or fog bank and is then scattered or reflected back to an observer's eye. Diffraction of the light bends the returning light rays slightly as they pass around the edges of water droplets. Interference among returning waves produces the light and dark bands that form the glory.

Glories appear as a full circle or nearly so, their colored rings centered on the observer's shadow. Their angular width (the angle across which your eyes move when looking from one side of an object to another) depends only on the droplet sizes, not on the distance to the cloud or fog bank. Typically, glories measure between five and twenty degrees across, with smaller droplets producing the larger glories. As with the corona, the more uniform the droplet-size distribution, the more distinct the rings appear. The inner glory generally shines the brightest, with duller rings surrounding it. When glory rings are colored, red always lies on the glory's outermost edge.

When German mountaineers in the early 19th century saw the glory, they attributed it to a haloed spirit beckoning them. They named it the "Brocken specter" after the Brocken, a prominent peak in Germany's Harz Mountains where these mountaineers first saw it. The Brocken specter legend says that one mountaineer became startled when confronted by the "haloed" human form hovering in the clouds before him. Some accounts suggest the vision so unnerved him that he lost his grip and fell to his death. Other versions report that he was drawn to approach the vision and thus fell to his death. In either case, he was ironically drawn to his death by a vision of himself.

Rainbows

Rainbows have been a part of our mythology and culture for millennia, from the story of Noah to Native American tribal beliefs, and from Buddhist traditions to Viking legends and leprechaun tales. Rainbows have also fascinated scientists, including Aristotle and Newton, throughout history.

We have considered optical wonders arising from light interactions with small water droplets, but now we'll look at the interaction with larger water drops—raindrops—that produces the most famous of optical wonders: the rainbow. The formation of a rainbow requires two components. The first is rain, or another source of airborne water drops, and the second is a strong light source. While sunlight produces most rainbows, a bright moon can also generate one, which is known as a "moonbow."

Sunlight striking falling raindrops is refracted from its straight-line path as it passes through them. The refraction not only changes the incoming sunbeam's direction, it also splits it into pure-color rays in the same way that a prism splits "white" light into the full spectrum. Some of these rays, now traveling along slightly different paths from the original beam, then reflect off the inside back surface of the raindrop. As the refracted and reflected light waves leave the raindrop, a second refraction spreads the colors into a parallel beam front that travels back toward the sun. Violet light emerges from the drop at an angle of 40 degrees relative to the incoming sunlight. Red light, at the other end of the spectrum, exits at 42 degrees. Intermediate colors emerge at angles between 40 and 42 degrees, forming the familiar banded rainbow arch.

Although we usually see the rainbow as a continuous spectrum of colors, only one color actually reaches our eyes from each drop. We see red light from higher-altitude drops, producing the outer bow band, while the other color rays leaving that drop pass outside our view. Violet light from lower-altitude raindrops generates the inner visible band. It therefore takes millions of falling raindrops to produce the full rainbow we see.

The Optimum Raindrop

In rainbows, size matters. The brightest rainbows appear with raindrop diameters between 0.3 and 1 millimeter. If raindrop diameters are greater than 1 millimeter,

Rainbows apologize for angry skies.

Sylvia Voirol

red, yellow, and orange colors brighten but blue weakens. With drops smaller than 0.3 millimeter, the reds weaken in favor of the blues. Very small droplets, like those in fog and mist, produce faint, almost white, rainbows. Not all light rays travel through the drop in this manner, however. Some are internally reflected off the drop's inner surface and reflected for a second time before leaving it at an angle of around 51 degrees. This produces a secondary rainbow outside the primary arch. The secondary rainbow is dimmer, about 43 percent less bright than the primary bow.

A double rainbow forms during a distant downpour. Note the outer bow has its colors reversed from the primary rainbow sequence.

The combination of primary and secondary bows is known as a double rainbow. While the primary rainbow has violet/blue on its inner edge and red on the outer, the secondary bow has its color sequence reversed: red on the inner edge and violet/blue on the outer. Between the two lies a dark region, known as "Alexander's dark band," where the returning light has been bent out of view. In contrast, the rainbow center appears much brighter than the surrounding sky because the incident sunlight has reflected directly back from the drop surfaces. We can always locate the rainbow's center by imagining a line projected from the sun, through our eye position, and onto the sky before us. This center point, called the "anti-solar point," rises and falls with the sun's elevation.

Unfortunately, no pot of gold lies at the rainbow's end because rainbows have no end; each actually forms a circle. How much of the circle we see depends on the sun's elevation. The most complete occur when the sun is on the horizon. As it rises in the sky, the anti-solar point lowers, and eventually the rainbow sinks completely below the horizon. For this reason, midday rainbows are an uncommon sight to ground-based observers.

R A I N B O W S

What they are:

Rainbows are brilliant arcs of color caused by sun- or moonlight striking falling raindrops. The color sequence of a rainbow has red on its outside band and violet/blue on the inner band. At times, two distinct bows may form, creating a double rainbow. The outer secondary bow has a reversed color sequence to the main bow. The rainbow always appears in the sky with the sun at the viewer's back.

What happens:

Light beams entering a raindrop are refracted on entry, reflected off the back surface of the drop, and refracted again upon exit from the drop as they head toward the viewer's eyes. The refractions split the incoming light into the full spectrum of colors. The full scene of the rainbow we see actually arises from the contributions of millions of raindrops.

The brightest rainbows appear with raindrop diameters between 0.3 and 1 mm. Double bows form when light reflected off the back of the drop is reflected again inside the drop. A dark band separates the two bows. The sky inside the rainbow arc is always bright.

Where to see them:

Rainbows appear just about anywhere in the world that has rainfall.

When to see them:

Rainbows can be seen in all seasons, though less frequently in winter at higher latitudes.

A schematic of how the different pathways of light rays through a raindrop form a double rainbow. Note the reversal of color between the primary and secondary bows—a feature that is clearly visible in the photograph on page 61.

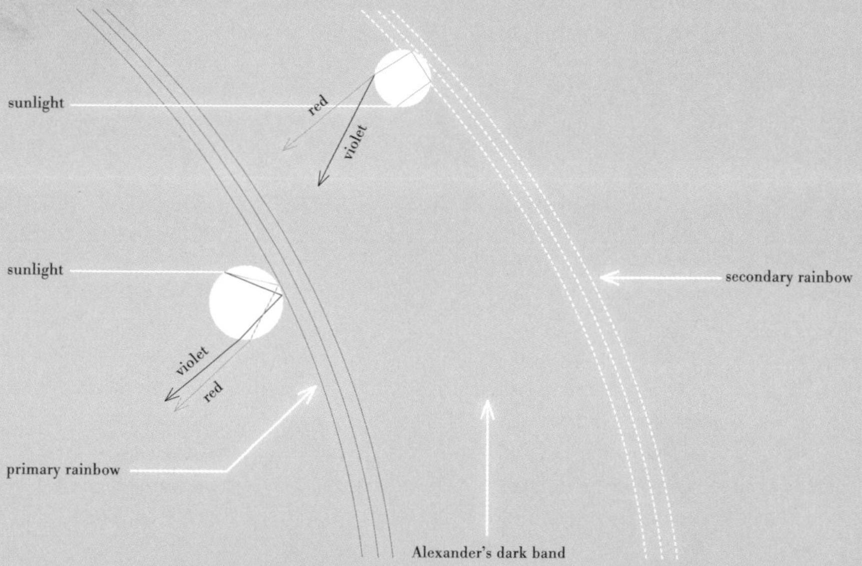

sunlight

red

violet

sunlight

violet

red

secondary rainbow

primary rainbow

Alexander's dark band

Ice-Crystal Phenomena

While water droplets and dust produce interesting and beautiful wonders in the sky, many spectacular optical phenomena arise from natural light shining through or off small ice crystals suspended in the air. These include many types of haloes, circles, arcs, bright spots, and pillars. Ice-crystal phenomena can rival rainbows for color and intensity, and several have served as indicators of coming weather for millennia.

Much of Earth's ice is held in oceanic and continental ice sheets, glaciers, and snow cover. A small fraction, however, floats in the air as individual ice crystals, usually congregated in high-atmosphere clouds. Interactions of light with these suspended ice crystals produce—through refraction, spectral splitting, and, at times, reflection—a variety of ice-crystal optical phenomena. Besides the basic halo, which encircles the Sun or Moon, these wonders include arcs and circles (or portions thereof), sun- and moondogs, and light pillars, plus rarely seen phenomena such as Parry arcs, anthelions, and a variety of tangent arcs.

When ice crystals make up thin high-altitude clouds such as the cirrus cloud family, they are thin enough to allow sunlight to pass through. Haloes have a close association with cirrus clouds, which in turn have a place in the sequence of weather events. As a consequence, they and other ice-crystal optical phenomena play a strong role in weather lore and forecasting. Their appearance has proved to be an invaluable tool in predicting the weather

THE CRYSTAL BASE

Depending on the air temperature and humidity under which they form, ice crystal structures cover a variety of hexagonal shapes. Stellar ice crystals take the well-known shape of a snowflake (more correctly a "snow crystal" as flakes are the conglomerate of many snow crystals). Others form as thin needles. However, the crystal shapes generally associated with haloes and similar phenomena are elongated columns and flat plates.

In crystal terminology, an ice crystal's two hexagonal faces are called "basal facets" and the six rectangular faces connecting the basal facets are called "prism facets." Column and plate crystals are similar except for their relative prism-facet thickness, which is long in columns and thin in plates. You can think of a plate as a squished column or a column as a drawn-out plate. Typically, the larger crystal dimension is 0.05 to 0.1 millimeters across.

Column crystals look like a common hexagonal pencil, their prism facets long compared to the width of the basal facets. Ice columns typically form when the air

Clouds decorate the sky with their jewellery of rainbows, halos, and coronas. They alter our moods and imbue our souls with profound emotional and symbolic content. In religious art, they serve as the footstools of the gods.

Stanley David Gedzleman

temperature falls into the ranges 23°F (–5°C) to 18°F (–8°C) and below –13°F (–25°C). Larger column crystals fall with their long axis parallel to the ground and, at times, rotate slowly like miniature helicopter blades.

Plate crystals are flatter, with the hexagonal basal facet width much larger than the prism facet depth, resembling microscopic stop signs or dinner plates. Favorable air-temperature conditions for plate-crystal formation fall in the ranges 32°F (0°C) to 25°F (–4°C) and 14°F (–10°C) to –4°F (–20°C). Very small hexagonal plates tumble randomly as they fall. With larger plates, aerodynamic drag forces orient the plates horizontally, and they fall with their large basal-facet surfaces parallel to the ground, descending like a large leaf drifting down from a tree.

HALOES

A halo—a brightly colored ring, or portion thereof—encircles the Sun or Moon when a cloud of ice crystals refracts its light in transit. The most common halo is the 22-degree halo, but on some rare occasions, a 46-degree halo may be seen outside it.

Haloes may also develop when light passes through the lower atmosphere containing ice fog, blowing snow, or "diamond dust"—suspended ice crystals in clear frigid air. For this reason, polar regions often display the most impressive haloes and related phenomena, particularly the Antarctic interior, where ice crystals tend to be homogenous in shape, size, and orientation.

To form a 22-degree halo, solar or lunar light rays must enter the ice crystals through one basal facet and exit through the other, being refracted as they do so. Light passing through these randomly oriented crystals—at the minimum deviation angle of approximately 22 degrees, with slight variation due to wavelength—is concentrated into a narrow band that makes up the halo's bright ring.

The 22-degree halo encircles the Sun or Moon either as a bright white ring or as a ring with some coloration. Typically, colored haloes have a distinct, red inner ring and a diffuse, bluish outer ring. Haloes rarely rival rainbows for distinct color banding because we actually see a number of overlapping haloes—each formed from an individual crystal—that compose the viewed halo. Now, because only light exiting at the minimum deviation angle forms the inner red ring, it cannot overlap other color wavelengths and therefore keeps its distinct color. The outer band, however, results from the overlapping of colored light rays from the contributing haloes, and this produces a white or bluish coloration.

The much less frequently observed 46-degree halo lies at a 46-degree angle from the light-source axis. It arises from the

same ice crystals as the 22-degree halo but forms when the light takes a different path through them. These rays go in one prism-facet side and then out of one of the basal ends. The 46-degree haloes have the same color sequence as the 22-degree.

Several hypotheses have been advanced to explain the rarity of the 46-degree halo. Some investigators argue that for it to occur the column ice crystal must be "fat." A 46-degree halo requires that light is refracted twice: once when it enters the crystal at the side, and again as it exits from the end. A long, thin column ice crystal makes it less probable that enough light will exit from its end, therefore preventing the phenomenon. Others suggest that the 46-degree halo's large width—it has an angular span of 92 degrees, compared to the 22-degree halo's width of 44 degrees—makes it hard to detect as large portions fall below the optical horizon. They may therefore be more common than we think.

One of the longest-standing sayings of weather lore cites the appearance of a halo around the sun or moon as a harbinger of stormy and wet weather. Richard Inwards' classic collection *Weather Lore* (1893) lists no less than 30 weather rules, sayings, or proverbs pertaining to haloes, many of which are generally correct in the mid-latitudes. Why are these weather sayings often true? In the typical cloud sequence preceding a storm system, an extensive band of icy cirrus clouds ushers in the warm front and its attendant precipitation. When sun or moonlight passes through these thin clouds, a halo often forms. Therefore, if the sequence runs true to form, the halo will presage the coming rain.

A full halo rings the Sun. Color differentiation is small in this example, but faint red is visible on the inner ring. Haloes form when light passes through ice crystals.

SUNDOGS

When sunlight passes through cirrostratus clouds, one or two bright spots may flank the shining orb. Technically called *parhelia*, meaning "with the Sun," we also call these spots "mock suns" or "sundogs." The former name acknowledges that in some situations, they can be as bright as the cloud-dimmed sun and thus be mistaken for it. The latter most likely recognizes their resemblance to dogs flanking their master. Sundogs rank second as the most frequently observed halo phenomenon behind the 22-degree halo, which they often accompany.

 Brilliant sundogs flank the solar disk over Manitoba, Canada. A portion of a faint 22-degree halo is also visible above them. Both sundogs and the Sun also exhibit sun pillars.

Both sundogs and haloes arise from light being refracted as it passes through the basal facets of column- or plate-shaped ice crystals. The difference in their appearance results from the crystal orientation. Haloes form from ice crystals randomly oriented in the cloud. However, if many horizontally oriented crystals are present, sundogs emerge. The more perfectly the falling crystals are aligned to the horizontal, the more compact and bright the resulting sundog. Misalignments from the horizontal spread the sundog. On occasion, sundogs sport bluish-white tails that stretch horizontally away from the Sun like a comet.

Sundogs appear bright white but may show some color banding with red wavelengths on the edge nearest to the Sun and blue on the outer. The degree of coloration depends on how much the ice crystals wobble as they fall: the more wobble, the greater the degree of color distinction.

Sun dogs and moon dogs are beautiful accents to a winter day or night as the rainbow is to a showery summer day.

Hal Borland, *Sundial of the Seasons*

Sundogs typically appear when the Sun is low to the horizon. At these elevations, not only cirrus clouds but other ice clouds such as ice fog and diamond dust may also generate them. If the Sun hangs low—from the horizon to about 15 degrees above it—each sundog sits at 22 degrees (about two handbreadths on extended arms) from the Sun but never closer, lying on the 22-degree halo if one is present. As the Sun climbs in the sky, sundogs slowly spread away from 22-degree separation although they always remain on a line through the solar center and parallel with the horizon. When the sun reaches 45-degree elevation, sundogs become faint and sit noticeably away from the 22-degree halo. They vanish altogether when the sun climbs above 61 degrees. Poleward from the mid-latitudes, sundogs may appear in the winter sky around midday when the Sun hangs low.

Thousands of years ago, the Greeks recognized that haloes and sundogs foretold rain. Today, we known this can be a valid forecasting tool, because they are produced by ice crystals in cirroform clouds, which make up the typical cloud sequence preceding a precipitating warm front. Sundogs are not the only "canine" optical wonders; we also have moondogs. These appear alongside the Moon, formed by lunar light passing through ice crystals. Moondogs (aka *paraselenae*) are not visible as often as sundogs because moonlight is only bright enough near the time of the full Moon to produce strong moondogs.

ARCS, TANGENTS, AND CIRCLES

At times, light passes through ice crystals along different paths from those forming the common haloes and produces other optical wonders: a geometry of arcs, tangents, and circles. The best, and most frequent, displays of many of these rare phenomena usually occur in the cold, clear skies of Antarctica.

The circumzenithal arc and the circumhorizontal arc, though not commonly seen, are among the most beautiful of ice-crystal optical wonders, having a brilliance and color that at times rivals the rainbow. As their names imply, these arcs occur in different parts of the sky. The circumzenithal arc inhabits the sky above the viewer—the zenith; the circumhorizontal arc lurks near the horizon.

The circumzenithal arc forms when light enters the crystal's upper basal facet and exits through a side prism facet of a large horizontally oriented plate. For the circumhorizontal arc, the light enters through a prism facet and exits from the lower basal facet. To set up the geometry of light rays and crystal refractions that produce these arcs, the Sun's position becomes important. For circumzenithal arcs, the Sun must lie below an elevation of 32.2 degrees from the horizon. For circumhorizontal arcs, the Sun elevation must exceed 57.8 degrees. These angle criteria prohibit the two arcs from being seen together and also limit when and where on Earth these arcs may appear. Circumhorizontal arcs cannot be seen at any time at

When I gazed up, a haze was spreading over the moon's face; and, as I watched, a system of luminous circles formed themselves gracefully around it. Almost instantly, the moon was wholly surrounded by concentric bands of color, and the effect was as if a rainbow had been looped around a huge silver coin.

Richard E. Byrd, *Alone*

locations poleward of 55 degrees latitude, and in some middle latitudes their appearance is limited to the summer months, often around the summer solstice. For example, at the latitude of London, England, the solar elevation only exceeds the minimum limit for a total of 140 hours.

Circumhorizontal arcs can form a complete ring but rarely do. The arc is found on the same side of the sky as the Sun, lying parallel to the horizon. It exhibits multicolored banding with red on the arc inside (sunside) band and blue on the outer band—a color sequence common to all halo phenomena and opposite to that of rainbows. Circumzenithal arcs never form a complete circle, and only rarely exceed a quarter circle. Some describe the circumzenithal arc as a colorful smile high above the solar orb. With upward curvature toward the zenith,

this arc has the common halo color-banding pattern with red on the Sun side (lower side). Circumzenithal arcs generally have a short life and faint to moderate brightness. The maximum brightness, width, and elevation of this arc occurs when the Sun elevation sits at 22 degrees.

Another arc with moderate observational frequency is an upper tangent arc to the 22-degree halo. This arc usually touches the top of the 22-degree halo directly above the Sun, looking like the frontal silhouette of a flying gull. With the sun low in the sky, the "wing" positions are more vertical, and they droop when the sun is high. Its color can be very bright and is usually white or reddish, although typical halo color banding is possible. Less frequently seen is the lower tangent arc—similar to the upper arc but lying inverted directly below the sun. A ground-based observer only sees this arc when the sun sits between 22 and 29 degrees above the horizon. When the sun rises higher in the sky (above 30 degrees) the two tangents reach out and join, forming a circumscribed halo. The tangent arcs form similarly to the sundog and 22-degree halo, but must have the long axes of column crystals aligned horizontally.

A Sun pillar rises above and below a low-elevation Sun, creating the impression that a light is shining from the ground up toward the heavens.

The "parhelic circle" appears as a white band circling the sky and always lies at the same height above the horizon as the Sun. This circle cuts the Sun, joining the 22-degree halo at the points where sundogs reside. It arises from both the external and internal reflection of light off the vertical facets of ice crystals. Internal reflections may be multiple reflections, following whatever path will send them back toward the viewer. While a parhelic circle formed only by surface reflections always appears white, some coloration may occur when various internal reflections produce a net refraction on exit. The parhelic circle occurs with about the same frequency as the 46-degree halo.

LIGHT PILLARS

When looking toward the rising or setting Sun, we may notice a shaft of light extending vertically above it like a bright feather. This shaft, known as a "light pillar," forms when the light reflects off airborne ice crystals. Although Sun pillars are the most common light pillar, light from the Moon may also produce them. Pillars can occur in all seasons but do so most frequently during winter, when ice crystals are nearer the ground. Most light pillars extend 5 to 10 degrees above the shining orb but they can appear below it as well. Because crystals reflect the light source, pillars show its color, often the oranges and reds of sunset or sunrise. With a higher Sun, pillars appear white.

Plate or column ice crystals provide excellent reflective surfaces. Cirrus clouds, ice fogs, diamond dust, and blowing snow all contain both crystal types. When these crystals congregate in a stable "flying formation" with one axis parallel to the ground, they can act like an array of small mirrors to passing light rays. The breadth and position of a light pillar depend upon the type and orientation of the crystals, their altitude and distance from the observer, and the elevation of the light source.

Plate crystals only produce Sun pillars when the Sun lies within 6 degrees of the horizon. Plates of roughly 1 millimeter across, when their long axes are aligned almost perfectly horizontally as they fall, form narrow, bright pillars. If the crystals deviate from perfect alignment, the resulting pillar broadens and may detach from the light source. Light reflected off ice columns can form pillars when the Sun is higher, though rarely when the sun sits more than 20 degrees above the horizon.

When the ice crystals are at a great distance from the observer, some slight horizontal misalignment of the crystals is required for pillars to appear. Light rays reflected off distant, perfectly horizontal plates do so away from the observer's line of sight. With crystals slightly tilted, the light can reflect off the lower outer crystal surface (or inner upper surface) toward the observer.

With ice crystals high in the sky, only reflections off their bottom surfaces form pillars visible to ground-based observers. When crystals float near the ground (and therefore the observer), however, as in the case of ice fog, blowing snow, or diamond dust, light may reflect off crystals above or below the direct viewing path, thereby producing pillars that extend both above and below the source. Crystals near the ground not only produce pillars with sun- or moonlight, but can also generate very interesting optical effects around strong artificial light sources.

LIGHT PILLARS

What they are:

As the name suggests, light pillars are stationary columns of light, appearing to hang in mid-air above and/or below a light source, such as the Sun. Their color can vary from orange to bright, bluish white.

What happens:

Light pillars are caused by light reflected off falling ice crystals. These ice crystals, greater than 0.04in (1mm) across, must have plate or column shapes descending with the correct orientation—a stable "flying formation" with one axis parallel to the ground. The breadth, form, and location of a light pillar depend upon the ice-crystal type and orientation, the height of the crystals in the sky and their distance from the observer, and the altitude of the light source. The Sun forms most light pillars, but they can appear from moonlight or even artificial lights. The pillar color varies depending on the source. Sun pillars are white when the Sun is high in the sky or predominantly yellow, orange, or red when it is near the horizon.

Where to see them:

You need two things to see a light pillar: a bright light source and ice crystals in cirrus clouds or near the ground, as in the case of blowing snow or diamond dust. Therefore most, but not all, sightings take place in cold countries. At times, they are associated with haloes.

When to see them:

Light pillars can occur in all seasons but are most frequent during winter. Pillars formed from artificial light are most commonly seen on cold, windless nights when ice crystals form and fall through the air with a horizontal orientation.

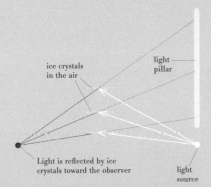

ice crystals in the air

light pillar

Light is reflected by ice crystals toward the observer

light source

A Sun pillar in Arizona.

The Curtains of Heaven:
The Aurora

High above the magnetic poles, the solar wind from the Sun
interacts with Earth's magnetic field and atmosphere to produce the
incredible beauty of the aurora borealis (Northern Lights) around
the northern magnetic pole and the aurora australis (Southern
Lights) around the southern pole. All the peoples of the polar
regions have myths surrounding the aurora. The word *aurora* comes
from the Roman goddess of dawn, Aurora, for its resemblance to
the first light of day.

Auroras are the offspring of high-energy particles blown off the Sun interacting with Earth's geomagnetic field to excite the tenuous oxygen and nitrogen atoms and molecules in the upper atmosphere, a process similar to that behind neon lights.

Auroras form within oval bands centered at each magnetic pole that extend outward approximately 1,875 miles (3,000km) during quieter solar periods. However, when solar storms disturb Earth's magnetic field and upper atmosphere, the auroral oval may extend much farther out from the magnetic pole, as far towards the equator as 35 to 40 degrees latitude.

Earth's magnetic field channels high-energy, solar particles (primarily electrons and hydrogen ions) to collide with neutral and ionized atoms or molecules of the outer atmosphere gases. The collisions force their electrons to absorb the particles' energy and jump to a higher, excited energy state. When excited electrons return to their initial energy state, they relinquish the absorbed energy as photons of colored light. Which color we see depends on the specific atmospheric gas being struck, its electrical state at the time of collision, and the energy of the extraterrestrial particles involved in the collision.

An aurora borealis dances over the winter sky at Bear Lake, Eielson Air Force Base, in Alaska. Auroras are caused by solar particles hitting the upper atmosphere, and their visibility is typically limited to the polar regions. In one famous event in 1859, however, the phenomenon was visible at much lower latitudes, including as far south as the island of Cuba.

[The aurora's serpentine motions] . . . were like a tai-chi exercise: graceful, inward turning, and protracted.

Barry Lopez, *Arctic Dreams*

Each excited atmospheric gas emits specific colors. The brightest and most common auroral color, a brilliant green-yellow, emanates from oxygen atoms at an altitude of roughly 60 miles (100km). Rare, all-red auroras originate from very high-altitude (about 200 miles/320km) oxygen atoms struck by extremely energetic solar particles. Ionized nitrogen molecules produce blue light while electrically neutral nitrogen molecules create purplish-red auroral colors.

Auroras may appear to hang static in the sky, or they may dance and flutter, and even explode upward along magnetic field lines. Auroras come in many shapes, sizes, and colors that vary as the solar wind buffets the geomagnetic field with solar particles. We define six common shape categories for auroral displays:

Auroral arcs or curtains typically cover large areas of sky. Usually an evenly curved arc of light with a smooth lower boundary, they may extend from horizon to horizon. Arcs may dance and turn, rippling like a curtain responding to a light breeze.

Auroral bands are similar to arcs but with an uneven lower edge. Bands and arcs have sub-categories: "streaming" when a sudden increase in brightness passes horizontally across them; "flaming" when sudden bursts of light begin at the bottom and flare upward; and "rayed" when composed of distinct rays.

Auroral coronas are rayed arcs or bands seen from below, appearing as sunbeams spraying out in all directions.

Auroral patches are small areas of light resembling a cloud in the night sky. Often they flash or pulse, disappearing and reappearing over a short time period.

Auroral rays are thin beams or shafts of auroral light hanging vertically above the observer. Tall rays generally start 60 miles (100 km) above the Earth and extend vertically along the magnetic field for hundreds of miles.

Auroral veils are large, featureless auroral clouds covering the entire sky with pale white light. In exceptional cases, veils may appear blood-red, and such vivid displays have inspired myths and legends among northern peoples, in many of which they portend disaster.

Many consider auroras a winter-night phenomenon. The longer dark nights and clearer skies common during the winter do make their faint light more visible, but as polar-orbiting satellites show, the auroral ring continually glows at each magnetic pole, waxing and waning with variations in the solar wind regardless of season. They dance even in daylight, but the sun's overwhelming brightness hides their faint glow from our sight.

Perhaps the association of the aurora with winter comes from the experiences of inhabitants of the far north, the Land of the Midnight Sun, where it is light all day during the summer and perpetually dark during the winter. These indigenous peoples have watched the aurora for millennia and developed strong myths and legends. Some see them as benevolent, others as evil, but all have supernatural explanations.

The Inuit and Eskimo peoples of polar North America told many stories about the aurora borealis, which differ between the tribal regions. Many believed auroras were the spirits of the deceased, including children lost during childbirth. Others saw auroral displays as dead souls playing "football" with a walrus head. In contrast, the Point Barrow Inuit of Alaska saw the lights as evil and carried a knife for protection. The Saami of northern Scandinavia also believed the lights emanated from the dead, and that therefore the aurora must be respected. To disrespect it risked bringing illness or death to the village; whistling at the aurora, they believed, would lead them to harm.

Magnificent aurora borealis, or Northern Lights, hang as curtains that waver in the sky over a church in Iceland.

75

Other cultures saw auroras as animals, including schools of herring, or the spirits of animal species they hunted. Chinese and Russian folklore saw auroras as a fire-breathing dragons. Some, such as the Maori of New Zealand, the Makah of Washington State, and the Mandan of North Dakota (U.S.) thought the lights resulted from supernatural fires.

The Vikings had many stories and legends based on the aurora, including that it came from heavenly light reflecting off the shields of the Valkyrie maidens who took dead warriors to Valhalla (heaven). Another held that it came from an erupting volcano in the far north provided by the gods to bring light and warmth to the polar regions.

For peoples outside the polar regions, where auroras were infrequently seen, their appearance often portended something bad. Children were brought indoors for fear the auroras would descend and take the youngsters' heads. In North America, the Fox Nation feared the light as being the ghosts of slain enemies waiting to take revenge. The Greeks believed they foretold coming wars and disease. Nordic peoples believed that the gods were angry when the aurora flamed. German tribes saw the auroral fires as Valkyries riding through the air, a scene later described musically by Richard Wagner in his dramatic "Ride of the Valkyries," the prelude to the second act in *Die Walküre*, the second opera in his Ring des Nibelungen cycle. The piece was also used in a scene in the movie *Apocalypse Now.*

Among the most remarkable auroral displays across the U.S. in the past century was an event that occurred on the night of March 7–8, 1918. It disturbed the sky as far southeast as Miami, Florida—where it caused the northern sky to turn red—and southwest as far as Texas and New Mexico. In central Illinois, H. Merrill Wills described it as "the most brilliant ever witnessed in this part of the country . . . the hues ranging from deep rose to purple, yellow, white, and green, extended upward from the main arch in sunburst effect."

Globally, the incredible dual auroral events of August 28 and September 1, 1859, mesmerized viewers across Europe south to Italy, in North America as far south as Cuba and Central America, and in Japan, Hawaii, Chile, and Australia. No other global event would be so extensively reported until the eruption of Krakatoa 24 years later. To the citizens of Havana, Cuba, the sky that night "appeared stained with blood and in a state of general conflagration."

The frequency of large magnetic storms that are a major cause of auroras follows the 11-year sunspot cycle, peaking within and remaining high for the following three years. Currently in 2010, we are coming out of the sunspot minimum of the current cycle. Therefore, we can look forward to increased auroral activity over the next few years.

AURORA

What they are:

Auroras are high atmospheric phenomena formed by the interaction of Earth's atmosphere with solar wind particles. Auroras may appear to hang static in the sky, they may dance and flutter, or even explode upward along magnetic field lines. Auroras appear in many shapes, sizes, and colors that vary as the solar wind buffets the Earth with solar particles. They may form as sheets, curtains, patches, rays, and veils, and are usually colored red, green, or white.

What happens:

Energetic solar particles caught in Earth's magnetic field collide with high-atmosphere gases, sending the gases' electrons to an excited state. When excited electrons return to their normal level, light is emitted, the color depending on the specific gas struck, its state at the time of collision, and the energy of the extraterrestrial particle.

Where to see them:

At latitudes around the magnetic poles, down to the high mid-latitudes, though on rare occasions they may appear throughout the middle latitudes.

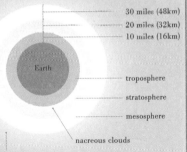

30 miles (48km)
20 miles (32km)
10 miles (16km)

Earth

troposphere

stratosphere

mesosphere

nacreous clouds

noctilucent clouds

Above: Very high-altitude clouds reside in the uppermost atmosphere. Nacreous clouds are found at heights of 10–20 miles (16–32km) while noctilucent clouds reside 45–55 miles (70–90km) up.

Below: Another spectacular display of the Northern Lights in the skies over Alaska.

When to see them:

At night. Though auroras are ever-present at each pole, the hemispheric winters are best for viewing due to the longer nights. In the summer, they disappear under the midnight sun but can be seen at night south of the polar circles. Auroras develop more frequently when the sun is active, such as around the peak in the sunspot cycle.

CHAPTER THREE

atmospheric

phenomena

Solar energy beams across Earth, warming the atmosphere, land, and oceans. To achieve a balanced distribution of the Sun's energy, a global atmospheric heat engine uses wind, clouds, and precipitation to move heat poleward. As we explore this remarkable process, various phenomena become evident.

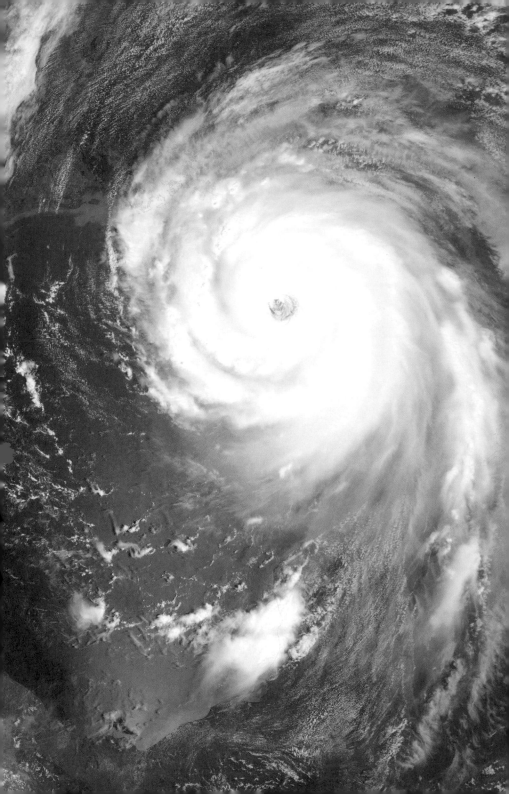

And How the Wind Doth Blow

Driven by solar energy, the atmosphere is never still. The wind, air in motion, blows around the globe in strong bands at high altitudes, in great swirls called cyclones, and in smaller eddies around and among the terrain. No wonder peoples around the world have credited winds with having supernatural personas, from the Greek Aeolus to the Hindu Vayu, and have personified winds as the *landlash*, the *kohala*, and the *matsuhaze*.

In the previous chapter, we looked at some of the optical wonders arising from the direct interaction of sunlight with air, water, and ice. The majority of that sunlight, about 70 percent, goes to heating the atmosphere, the oceans, and the land, converting the solar energy into heat. The greatest heating occurs in the tropical regions, where the solar beam is most direct, while the least direct occurs at the poles. We know that the equatorial regions are hot and the poles cold, and this condition cannot be sustained. Various processes attempt to produce a temperature equilibrium across the planet. While this equilibrium never happens, it is not for a lack of effort on the part of the earth. Two major systems, the great ocean currents and the global wind field, continually move heat poleward from the tropics.

Earth scientists call the overall mechanism that transports this heat the "global heat engine," and it must be considered as a natural wonder, dwarfing the greatest of industrial engines in size and complexity. In this section, our focus is on the atmospheric portion manifested through the global wind field. Unfortunately, we only see the winds at work indirectly as they push clouds around the planet. By stepping away from the planet and using time-lapse movies of global satellite images, we can see moving masses of clouds swirl around the planet, sometimes in great whirls known as cyclones, which in the tropics may develop as smaller but more violent storms known as tropical cyclones, hurricanes, and typhoons. Within the larger swirls, smaller violent whirls can arise. Known as tornadoes, they blow with wind at speeds faster than the strongest hurricane.

This image, captured by the Moderate Resolution Imaging Spectroradiometer (MODIS), shows Hurricane Katrina on August 28, 2008, at 13:00 EDT. The massive storm covers much of the Gulf of Mexico, stretching from the U.S. Gulf Coast to the Yucatan Peninsula. The eye of the storm is clearly visible.

Did you ever take pencil and book to scribe down the sounds the wind makes as it sifts and soughs through the trees?

Guy Murchie, *Song of the Sky*, 1954

There is no doubt that hurricanes, tornadoes, and the strongest of non-tropical cyclones, such those that manifest themselves as winter blizzards, are wonders of power and horror when they rage over human habitations. These fits of nature contain the power to easily break and uproot trees, down utility poles, and reduce houses and other buildings to rubble. Our focus here, however, is on winds of a local nature that arise within the global wind field.

LOCAL WINDS

By local winds, we mean those that act primarily on a local scale, which are smaller than continental winds but bigger than the gusts and eddies that swirl around trees and buildings. Some such winds are adjuncts to the larger continental-scale winds that continually blow across the planet, and arise when these winds encounter specific types of terrain. The first example we shall examine is a wind that is known as the *foehn* in Europe and *chinook* in North America. The second are dust- or sandstorm winds that go by many names, including the *haboob*.

Local winds also arise from unequal heating of the land surface by sunlight. Unequal heating results in the formation of small low-pressure cells of hotter air surrounded by higher pressure cooler air. Wind is air flowing down a pressure gradient (a change in surface pressure with distance) from high to low pressure, and on the local scale it flows from cooler air toward warmer air.

The most common examples of such winds are the sea and land breezes that often blow along coastal areas of large lakes, seas, and oceans. Sea breezes flow inland during the day when the land is heated by the Sun, establishing a flow from water to land (cold to hot at the surface). At night, when the air over the land cools faster than that over the water, a reverse flow, known as a "land breeze," moves land air offshore and over the water body. In some areas of the world, a cool sea or land breeze can replace stifling hot, humid air, providing relief to the local human population. Locals often refer to these winds as "the Doctor."

In situations such as the sea breeze, the air flows predominantly horizontally, but in some cases the heated air rises vertically instead and, while rising, the air is given a twist. We call this family of local winds "devils," led by the dust devil. The aquatic member of the devil family is the waterspout, although its most violent manifestations are more closely related to the tornado.

FOEHN AND CHINOOK WINDS

When the global-scale winds, such as the Westerlies that blow from west to east in the mid latitudes (30 to 60 degrees latitude), flow over high mountain ranges—the European Alps or the American Rockies, for example—they descend the leeward slopes after crossing the ridge line. As they descend, they warm through compression, becoming hotter and drier than the surrounding air. First studied in the Alps, these winds are generically known by their local Alpine name: the foehn winds.

For an example of how foehn winds occur, let's look at the North American situation downwind (that is, to the east) of the Canadian Rockies in winter. The Westerlies blow across the continent from the Pacific Ocean to the Atlantic shore. We first meet them approaching the Pacific coastline having crossed Pacific Ocean waters, where they have picked up heat and moisture from the ocean.

When these relatively warm and very moist winds make landfall along the British Columbia coast, they encounter the coastal mountains and rise over the terrain as they trek eastward. While crossing these mountains, the winds cool as they rise, forming clouds and then precipitation, which eventually falls in prodigious amounts of rain or snow that waters the lush, temperate rainforests for which the region is famous.

Although this air stream cools and expands as it rises over the mountains, it gains back some heat when its water vapor converts to liquid water through the latent heat of condensation. (Even more heat can be gained if liquid water freezes, releasing the latent heat of fusion.) By the time the air has traversed British Columbia's parallel mountain ranges, much of its water content has been lost through precipitation; however, a large portion of that released latent heat remains in the air.

When the airflow finally begins to descend from the lofty ridges of the Rockies onto Alberta's high plains, it heats up as the air compresses—like air in a bicycle pump, which heats up when we push the plunger down—while descending to lower altitudes. The warming amounts to about 5.4 Fahrenheit degrees per 1,000 feet of descent

(9.8 Celsius degrees per 1,000 meters. Since many ridge lines in the Canadian Rockies reach 10,000 feet (3,000m) above sea level and the Alberta plain is around 3,300 feet (1,100m), the descending air warms around 36 Fahrenheit degrees (20 Celsius degrees) in its descent. The air parcel is now also very dry, since it has lost most of its initial moisture content crossing the mountain chains. This warm descending air is known locally as the *chinook*.

The most impressive chinook winds blow off the Rockies at speeds of between 40 and 60mph (65–96km/h) with gusts exceeding 100mph (160km/h). At those speeds, chinooks can tip railcars off the tracks. Impressive as the chinook winds are, the temperature changes they bring can be more astonishing, often as much as 36–45 Fahrenheit degrees (20–25 Celsius degrees) in an hour. The greatest chinook temperature jump ever recorded occurred on January 22, 1943, when the temperature in Spearfish, South Dakota, shot from a chilling −4°F (−15°C) at 7:30am to 47°F (8°C) just two minutes later!

In winter, the combination of relatively hot and dry air rushing by at high speeds can remove a foot (30cm) of snow in a few hours, often turning most of it back to vapor. In other seasons, foehn winds can raise soil into dust storms, quickly desiccate vegetation, and rapidly increase the danger of grass and forest fires. Many who live under the foehn's influence suffer debilitating physical effects, ranging from sleeplessness to anxiety and severe migraine headaches. Foehn winds generally last only a few hours but can continue for several days, sometimes even persisting for several weeks. They go by many different names around the world. In North America, they have several other names including the feared Santa Ana winds of California. In Libya, their local name is *ghibli*, while in Java they're called the *koembang*. Down the Andes, the *zonda* blows across the Argentinean pampas. In Japan, they call the foehn the *yama oroshi* ("storm blowing down from the mountain").

Haboobs and Other Dust Winds

When the winds blow across land where sand or dirt dominates the surface cover, that surface material can be raised into the air. "Aeolian transport" (after Aeolus, a Greek god of the wind) is the term used to describe particles carried by the wind and held aloft through suspension. The Aeolian process usually starts when strong, gusty winds blow across arid landscapes and loft small particles such as sand, dust, and soil into the sky.

While any sufficiently strong wind may raise dust or sand into the air, most local-scale, airborne dust events arise during the passage of weather fronts (where contrasting air masses clash) or thunderstorms.

Occasionally, blowing dust covers a very large area; this is termed a dust storm. These dust storms form under sustained, strong surface winds associated with larger-scale windstorms. Dust storms can last from three or four hours up to two or three days. Dust storms carry colorful local names such as the *haboob* of the Sudan, the *sirocco* of North Africa, the *simoom* of Arabia, the *harmattan* of Algeria, and the *khamsin* of Egypt. Northern Africa, however, does not have sole claim on interesting dust-storm names. In North America, they are called "black rollers" or "black blizzards"; in Argentina, the *zonda* carries dust across the pampas; in Australia, it's the "brickfielder."

Leaping and Creeping

Wind moves surface particles in a variety of ways: suspension, saltation, and creep. Suspension carries dust particles lofted into the air and holds them aloft in air currents. Typical surface wind speeds, blowing around 10mph (16km/h), can suspend dust particles with diameters less than 0.008in (0.2mm) and carry them short distances. Severe wind-

Who has seen the wind?

Neither I nor you.

But when the leaves hang trembling,

The wind is passing through.

Who has seen the wind?

Neither you nor I.

But when the trees bow down their heads,

The wind is passing by.

Christina Rossetti, "Who Has Seen the Wind?"

storms, however, can hold large particles aloft for some time by pushing them to very high altitudes. Thus, dust and sand can travel long distances, even across oceans.

In saltation, from the Latin word for "leaping"—an apt description of what dust and sand particles do at high wind speeds—particles advance through a series of jumps or skips, like a game of progressive leapfrog. The wind initially lifts the particles 3 to 7 feet (1–2 m) into the air. Those too heavy to remain suspended drift downwind a distance of approximately four times the height they had attained above the ground (i.e., one foot up equates to four feet downwind). When these return to earth, many hit other particles, causing them to leap up and advance.

Often both the original and the struck particle leap from the ground and advance downwind. In a chain reaction, the saltation process continues as long as wind speeds remain high. If a leap becomes sufficiently high, the particle may become suspended and pushed higher and farther along. When viewed from a distance, a field of saltating particles appears as a fuzzy layer next to the ground.

The saltation and suspension processes are dependent upon the ability of wind to lift and move particles. This ability, which also depends on the size and density of the particles, increases proportionally to the wind speed. If the wind speed doubles, it is

This photograph from 1935 shows a black roller, or black blizzard, descending upon a farmstead in Stratford, Texas, during the Great American Dust Bowl of the 1930s

eight times more effective in moving particles. If it triples, it is 27 times more effective. In this way, the ability of wind to carry such particles is expressed as a cube of wind speed.

When saltating particles strike heavy particles, they may only nudge them along—a slow sliding and rolling movement known as "creep." Creep usually only requires wind speeds exceeding 10mph (16km/h) to impart enough energy for a saltating particle to move a larger one.

If lofted dust forms a large cloud that sweeps forward with the wind, it creates a dust storm. Atmospheric situations causing dust storms generally fall into two main categories: localized convective events, generally associated with thunderstorms, and large-scale, non-convective events, such as weather fronts. Convective-event dust storms arise when strong, gusty winds blow out of thunderstorms and hit the ground, raising dust and sand along their path. The duration of dust storms depends on the lifetime of the process causing them—from a few seconds to many hours. They are most common during the late afternoon in spring and summer when daytime heating is highest.

Large-scale, non-convective dust-storm events develop under sustained high surface winds associated with cyclonic windstorms. Such dust storms can last up to several days.

These dust storms most often occur in late winter or early spring when extreme continental pressure gradients produce high wind speeds and surfaces are dry and bare. Approximately two thirds of all dust storms are associated with the passage of weather fronts and low-pressure troughs.

DUST AND OTHER DEVILS

Shifting from horizontal to vertically dominant wind events, we stay with the dust theme and look at the world of the various "devils." A natural wonder of everyday weather, an estimated ten dust devils roam Earth at any given moment. Actually, devils need not carry dust; these swirling winds can carry aloft sand, snow, leaves, and other debris.

Dust devils have many names around the world, including "whirlwinds," "dancing dervishes," and "desert devils." In California's Death Valley, they are called "sand augers" or "dust whirls." The Navaho people of the American Southwest named them *chiindii*, believing them to be ghosts of the dead. They considered clockwise-spinning chiindii as good spirits while those spinning counterclockwise were bad spirits. Australians label them "willy-willy," "whirly-whirly," or "Cockeyed Bob," terms probably derived from Aboriginal words. Egyptian dust devils are *fasset el 'afreet* ("ghost's winds"). To the

Dust storms in three shapes. The whirl. The column. The sheet. In the first the horizon is lost. In the second you are surrounded by 'waltzing Ginns.' The third, the sheet, is 'copper-tinted.' Nature seems to be on fire.

Michael Ondaatje, *The English Patient*

Kikuyu of Kenya, they are known as *ngoma cia aka* ("women's devil").

Technically a dust devil is a vortex—a circular, closed flow where the air circles an axis of rotation. Whirlpools, tornadoes, and waterspouts are all vortices. In their largest, most impressive forms, we can confuse dust devils with small tornadoes. Both have rising shafts of air that swirl dust and objects around their rotating columns to make them visible. While dust devils show a similar structure to tornadoes, they have very major differences. Dust devils develop from the ground up through local heating and rarely connect to an overhead cloud. True tornadoes always descend from large, energetic cumulonimbus clouds, the product of larger-scale forces.

We generally think of dust devils arising over large arid expanses, a belief utilized by filmmakers to allude to arid conditions. However, even parking lots can generate enough surface heating to form a sizable devil, which often goes unnoticed due to the scarcity of dust to make the rising column visible.

Most devils arise on hot, sunny days over dry terrain when columns of hot air rise from the ground high into the atmosphere. Under intense solar heating, the surface temperature soars, transferring heat to the overlying air. The resulting hot air quickly becomes less dense than the surrounding air, generating a rapidly rising air column. As this air column ascends, something, perhaps the breeze aloft or a passing car at ground level, gives it a spin. As more hot air rushes into the rising vortex at the surface, the spinning intensifies. If the updraft column rises very rapidly, it may stretch higher, thereby tightening the vortex circulation and causing it to rotate even faster, like a spinning figure skater bringing in her arms. If the vortex's invisible winds pick up loose surface materials, a visible devil is born.

As long as hot air remains near the base of the devil, the circulation continues. Meanwhile, the surrounding wind field pushes the devil forward to dance across the terrain. However, once the hot air is depleted or the circulation balance broken— such as by contact with a large obstacle or terrain feature—the devil disappears.

Devils arise from individual rising columns, and therefore pairs or swarms of devils may skip across the landscape. Because they are small and relatively short-lived, they are not influenced by Coriolis forces arising from the Earth's rotation, and therefore devils may rotate either clockwise or counter-clockwise depending on the spin-initiating force. A pair of devils may spin with opposing rotations as they dance side by side across the landscape.

Most devils are small, usually less than three feet (1m) in diameter and less than 100 feet (30m) tall, and they are rather short-lived, not lasting more than a few minutes. In all likelihood, more devils form than are seen, as many arise over surfaces with no loose materials to gather into the vortex. Larger devils, however, garner much attention. The largest may grow to 1,000 feet (300m) wide and hundreds of feet high with a life span ranging from fifteen minutes to an hour or more. Average dust-devil winds swirl at 15–25mph (24–40km/h), but strong devils have been clocked at 95mph (153 km/h). They can amble along the countryside at speeds of 10mph (16km/h) or race across the terrain as fast as 60mph (100km/h).

A typical desert devil carries around 220 pounds (100kg) of dust, sand, or soil

The sea breezes do commonly rise in the morning about nine o'clock; they first approach the shore, so gently, as if they were afraid to come near it, and of times they make a halt and seem ready to retire.

William Dampier, pirate and author of *A Voyage Round the World*

aloft, while large devils can suspend enough material to fill six dump trucks. Although most contain dust, soil, and sand, any light material, such as straw or leaves, can be swirled into their circulation. Large dust devils can cause damage with their gusty winds and swirling materials.

WATERSPOUTS

Waterspout vortices bridge the gap between devils and tornadoes, and can be examples of either. They hold a prominent place in maritime weather lore. When a waterspout approached, European sailors would often cross silverware, hoping to dissipate the storm or ward off its striking. The Chinese believed that battling air dragons caused waterspouts. The Winnebago tribe of Wisconsin considered waterspouts on Lake Mendota to be the hero-god, Winnebozho, a "sea serpent," taking a shower bath. Although Winnebozho was generally good-natured, he loved playing pranks such as overturning canoes with his body or tail.

We define the waterspout as an intense vortex funnel, sometimes destructive, of

The funnel-shaped column of a dying waterspout extends down from dark skies, disturbing the waters at Green Turtle Cay, The Bahamas.

small horizontal extent occurring over a body of water. Waterspouts are convective vortices like tornadoes, dust devils, and landspouts, whose circulations are driven by, or associated with, convection. A true waterspout forms over the water and will dissipate almost immediately upon touching land. At times, tornadic waterspouts rage, spawned by severe thunderstorms. These are true tornadoes that touch down on, or move across, a body of water.

Waterspouts commonly occur during the summer in very warm, subtropical waters such as those off the Florida Keys and Australia's East Coast. They can also be found around large lakes such as the American Great Lakes and Great Salt Lake, and Peru's Lake Titicaca during the late summer/early autumn when the lake surface waters are warmest and much cooler air crosses aloft. Fall waterspouts have been observed in the straits around Vancouver Island, British Columbia, and in the waters of Washington State's Puget Sound. In European waters, waterspouts are commonly seen across the western coast as well as the British Isles and several areas of the Mediterranean and Baltic seas.

True waterspouts form over the water away from any organized storm. They commonly develop beneath lines of rapidly growing, but shallow, cumulus congestus clouds, between 12,000 and 14,000 feet (3,660–4,270m) deep with bases about

2,000 feet (600m) above the surface. The waterspout funnel develops at the water surface and builds skyward. While the thin-columned waterspouts appear to be sucking water up from the water surface, we are actually seeing water vapor condensing in the rotating vortex of air. At the funnel base, the waterspout stirs the water into mushroom-shaped water sprays.

Waterspouts exist for only 15 to 20 minutes on average, with few lasting more than a half hour. They can move at speeds as high as 80mph (130km/h), but most amble slowly across the water. Like devils, waterspouts are not affected by the Coriolis force, and therefore they may spin either clockwise or counter-clockwise. Most waterspouts are 30 to 300 feet (10–100m) in height, but

they can extend to over 2,000 feet (600m). They measure a few feet to a hundred feet (30m) in width. Waterspout winds generally reach 60–120mph (100–200km/h), peaking around 140mph (225km/h). Like dust devils, waterspouts often form in families, as many as nine under the same cloud line. Ships at sea have reported as many as 30 in one day. An impressive outbreak raged across the Great Lakes during the seven-day period of September 27 to October 3, 2003, when over 66 waterspouts were sighted, including a single-day swarm of 21 waterspouts over Lake Ontario.

Unlike tornadoes, true waterspouts are not well defined and are considerably less destructive. However, waterspouts can be dangerous for boaters and near-shoreline structures, and are capable of overturning smaller boats and floats. However, their slow movement and high visibility allow most ships to steer clear. Waterspouts have struck some Gulf of Mexico oil-drilling platforms, but damage has always been minimal.

A waterspout off the Florida Keys photographed from a NOAA aircraft. Note the water spray at the surface where the spout touches the sea.

WATERSPOUTS

What they are:

Intense vortex funnels, sometimes destructive, of small horizontal extent occurring over a body of water. Waterspouts are convective vortices, like dust devils and tornadoes, whose circulations are driven by convection forming rising shafts of air. True waterspouts form over the water and will dissipate almost immediately upon touching land. Tornadic waterspouts are tornadoes spawned by severe thunderstorms, which touch down and cross a water body.

What happens:

True waterspouts develop beneath lines of rapidly growing, shallow cumulus congestus clouds. A waterspout develops when a shaft of warm air forms at the water surface, rises rapidly, and is given a spin. The waterspout funnel develops at the water surface and builds skyward. Though waterspouts appear to be sucking water from the water surface, it is water vapor condensing in the rotating vortex that actually makes the swirling funnel visible. At the funnel base, the waterspout stirs the water into mushroom-shaped water sprays. True waterspouts may spin clockwise or counter-clockwise.

Where to see them:

Most waterspouts form in very warm subtropical waters but they may also form over large lakes and along the coastline when surface waters are much warmer than the cooler air aloft crossing the water.

When to see them:

During the summer in subtropical waters and during the late summer or fall over lakes and northern coastlines.

Warm water is key to waterspout formation. While they can occur throughout the year, waterspouts are most common in the Florida Keys from June through September. It is not uncommon for multiple waterspouts to occur in a single day.

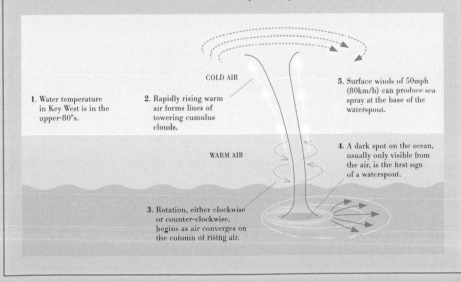

COLD AIR

WARM AIR

1. Water temperature in Key West is in the upper-80°s.

2. Rapidly rising warm air forms lines of towering cumulus clouds.

3. Rotation, either clockwise or counter-clockwise, begins as air converges on the column of rising air.

4. A dark spot on the ocean, usually only visible from the air, is the first sign of a waterspout.

5. Surface winds of 50mph (80km/h) can produce sea spray at the base of the waterspout.

Clouds

Who has not gazed at the sky and seen spectacular clouds that inspire us? They show gossamer filaments and robust masses, rounded like cauliflowers or towering like angry giants above the ground. High above the surface they shine with eerie light and mother of pearl. Clouds tell us of coming weather and warn of storms, while defining our planet's face in the solar system.

Viewed from space, Earth appears predominantly blue, a coloration resulting from its wide oceans. However, Earth shines not only ocean-blue with some green and brown continental patches, it also sports a white marbling, ever-changing bands that roam above the surface. These are the great systems of clouds, composed mostly of water in all its states: vapor, liquid, and solid (ice). On occasions, clouds of other compositions travel through the atmosphere: dust and sand, volcanic ash, and wildfire smoke and soot.

CLOUD FORMATION

Clouds have practical jobs in the atmosphere, adjusting the gain and loss of heat, moving heat and water around the planet, and providing precipitation that waters the great forests and grasslands. They also provide elements of great beauty that have inspired artists and poets for centuries.

When it comes down to the nitty-gritty, however, the recipe for clouds is rather simple: water with a pinch of dust. The first required ingredient, water in its vapor state, will be transformed into a liquid or solid. Next, we need dust—not a lot, and very small particles work best. Without dust, we would have no clouds, as it is required for condensation nuclei—sites on which water may condense or ice deposit. Certain particle types and shapes make the best condensation nuclei, such as sea salts and certain clays. The final step in cloud formation takes an air parcel containing these ingredients and cools it to a temperature at which cloud droplets or ice crystals form. Cooling usually requires ascent and expansion. Any of four processes occurring in the lower atmosphere can make air ascend: convergence, convection, frontal lifting, and physical lifting.

The large cumulonimbus cloud pictured here exhibits a fuzzy cirrus-topping anvil. These towering clouds are the product of unsettled atmospheric conditions and can develop into fearsome thunderstorms known as "supercells." They are responsible for the thousands of storms that rage across Earth every minute.

Cumulus nurslings emerge from their hot-air eggs
over warm fields and towns often only to blow along on the
wind for a few minutes to cooler regions, where mixing
with drier air, they fade and die.

Guy Murchie, *Song of the Sky*, 1954

Convergence occurs when horizontal air currents come together in a common space. When they converge at the surface, the only way to go is up. Convection occurs when air heated from below becomes less dense than the air above it. The heated parcel ascends until it has again cooled to the surrounding air temperature. Frontal lifting occurs where a warmer air mass meets a colder one. Since warm air is less dense than cold, it ascends over the cold air, forming a warm front. When cold air approaches warm, it wedges under the warmer air, lifting it above the ground, producing a cold front. In either case, air ascends along the frontal boundary. Physical lifting, or orographic lifting, occurs when topographical barriers such as coastlines, hills, and mountains force horizontal winds to rise in passing.

Whichever process causes air to ascend, the rising air parcel must reduce its pressure to be in equilibrium with the surrounding air. As air ascends, it expands, and as it expands, it cools. The higher the parcel rises, the cooler it becomes. As air cools, its saturation threshold decreases and its relative humidity—the ratio of water vapor present relative to its saturation threshold—will increase. As yet, nothing has changed the parcel's water-vapor content.

To form a cloud by changing the vapor to liquid, the air parcel's relative humidity

must be brought to saturation (100 percent) and a little—a few tenths of a percent—beyond, a state known as "supersaturation." When air becomes supersaturated, its water vapor begins to condense onto the condensation nuclei. Water molecules attach to the particles, forming cloud droplets with radii of 0.008in (0.2mm) or less. Cloud droplet volumes are generally a million times greater than the typical condensation nuclei. Condensation begins at a specific altitude, known as the "cloud base" or "lifting condensation level," which depends on the physical properties of the air mass through which the air parcel has ascended.

Billions of cloud droplets and/or ice crystals produce a cloud. Because of their small size and relatively high air resistance, droplets (or crystals) remain suspended in the air for a long time, particularly if they reside in ascending air currents. Average cloud droplets have terminal fall velocities of 0.5 inches (1.2cm) per second in still air. Therefore, the average cloud droplet falling from a typical low-cloud base of 1,500 feet (430m) would take 10 hours to reach the ground.

CLOUD TYPES

Prior to the 19th century, observers believed that clouds were too transient, too changeable, and too short-lived to be

classified. With few exceptions, no cloud types were given names but were described by their color and form: dark, white, gray, black, mare's tails, mackerel scale, woolly fleece, towers and castles. Englishman Luke Howard, in 1803, assembled the first widely accepted categorization of clouds. Howard believed that clouds could be identified by four simple categories:

Cumulus: Latin for "heap."
Stratus: Latin for "layer."
Cirrus: Latin for "curl of hair."
Nimbus: Latin for "rain cloud."

Howard observed that clouds could alter their shape from these basic categories, taking compound shapes such as when cumulus clouds crowd the sky into a layer. He called these "Cumulo-stratus," although today we reverse it and call them stratocumulus. The international meteorological

community accepted Howard's classification system in 1894 for standard observational use, adding additional terms, such as alto- (meaning "high") to designate the higher altitude where such clouds formed. Like Linnaeus's biological naming system, the simple groups or "families" of Howard's scheme have expanded to include "genus," "species," and, in some instances, "variety." For example, a cloud in the family *Cumulus* may be of the genus *Cumulonimbus*, which can be further described as species *Cumulonimbus capillatus*.

Dark storm clouds move onto the shore, likely bringing rain to Parker River National Wildlife Refuge, Plum Island, Massachusetts.

95

CLOUD TYPES

The following list defines the basic members of the cloud menagerie, divided into groups depending on their typical cloud-base height.

Low-base clouds with base heights up to 9,800 feet (3,000m).	**Stratus:** Clouds that usually appear as a uniform, gray, overcast deck but can occur as scattered patches. Individual cloud elements have very ill-defined edges.
Low-base clouds of large vertical extent with base heights of 1,600 to 9,800 feet (500–3,000m).	**Cumulus humilis:** Small heap clouds with flat bottoms and slightly rounded tops. Often regularly spaced across the sky—an indication of fair weather.
Middle-base (alto) clouds with base heights from 9,800 to 23,000 feet (3,000–7,000m).	**Altostratus:** These clouds differ from stratus by their elevated base height. They may have a bluish or gray hue.
High-base clouds with base heights above 23,000 feet (7,000m).	**Cirrus:** Clouds composed of ice crystals forming delicate filaments, patches, or bands.

Nimbostratus: Darker stratus clouds associated with areas of continuous precipitation. Bases are usually low. However, nimbostratus often extend vertically into the middle cloud region.

 Stratocumulus: Relatively flat clouds merged in tight clusters with little vertical development. Edges are more ill-defined than cumulus but more distinct than stratus. They may have a bluish or bluish-gray hue.

 Cumulus congestus: Heap clouds of large vertical and horizontal extent, often resembling a cauliflower. Bottoms flat and sharp, tops rounded with sharp outline.

 Cumulonimbus: The tallest of clouds; may extend to over 60,000 feet (1,800m). They resemble mountains or high towers and usually have large, fuzzy, anvil-shaped tops containing ice crystals.

 Altocumulus: Distinct cloud elements appearing either as a sheet or as a patchy deck with waved bands, rolls, or rounded masses—usually sharply outlined.

Cirrostratus: Ice clouds covering part or all of the sky with a transparent cloud veil of fibrous or smooth texture.

 Cirrocumulus: Ice clouds composed of small elements in the form of separated patches, ripples, and fish scales. They are more or less regularly arranged.

I am the daughter of Earth and Water,
And the nursling of the Sky;
I pass through the pores of the ocean and shores;
I change, but I cannot die.

Percy Bysshe Shelley, *The Cloud*

Other, less common cloud forms occur as well, many in special environments such as around mountain tops or very high in the atmosphere. Several of these are discussed below.

CLOUD COLORS

The basic elements of a cloud—water droplets and ice crystals—are transparent and so have no color. But when congregated into a cloud mass, they may take on many different colors and express many different moods. The cloud color and brightness depend on the angle of the sun or moonlight striking the cloud, as well as its primary color, the nature of the cloud, including its composition and thickness, and the viewer's position with respect to both the cloud and the light.

A child asked to paint a cloud will usually choose white, or perhaps black if painting a rain cloud. But in its basic natural form, a water or ice cloud varies in color from white to black through various shades of blue and gray. The reason for such coloration harks back to our discussion of sky color. The sunlight passing through a cloud is scattered and reflected by the droplets or crystals. Since they scatter wavelengths of all colors, the resulting light we see contains all colors and is perceived as white when clouds are thin and grayish blue as their vertical extent thickens.

In addition to scattering light, the cloud elements also reflect it. Droplets and crystals effectively reflect light but their large forms—raindrops and snowflakes—provide even greater reflective surfaces. This, in part, is the reason we see dark clouds as bringing rain or snow. When clouds are viewed from below, the raindrops (or snowflakes) have reflected much of the incoming light backward, and thus the cloud appears dark gray, even black, as often happens with large cumulonimbus.

Around twilight, clouds take on the colors of the setting or rising sun, enhanced by the cloud texture—a little more reflection here, a little less elsewhere—giving never-ending variations of beautiful patterns and colors. At times, low cloud bases take on hues derived from the surface color below them as the sunlight is reflected off that surface to the cloud base and then reflected to our eyes. Similarly, portions of a cloud may be illuminated by scattered skylight and thus take on a blue tint or even a pink hue. Cumuli may show "silver linings" when backlit because the thinness of the cloud edge is able to scatter forward more light than does the thick body, which extinguishes much of the light by reflecting or scattering it away from view. The reverse occurs when the sun shines from behind us. Then, the cloud body glows intensely bright white, due to reflection, while its edges appear dark.

Cloud color has long been used as a forecast tool in weather proverbs. In fact, some are so engrained they have become instinctual. If clouds are dark gray or black, we immediately suspect precipitation will fall shortly. Here are a few examples of cloud-color proverbs:

Red clouds in the east; rain the next day.
Anon.

Red clouds at sunrise foretell wind; at sunset, a fine day for the morrow.
Francis Bacon

If clouds rise in heaps of white, soon will the country of the corn priests be pierced with arrows of rain.
Zuñi Tribe of Southwestern USA

CAPPING THE MOUNTAIN SUMMIT

One special cloud formation not included in the basic cloud atlas is the cap cloud family, which, as the name denotes, looks like a cap covering a mountain peak. In addition to the basic shape, the cap cloud family includes the banner cloud, the rotor cloud, and the often-spectacular lenticular cloud. Typically, cap clouds sit on or very near mountain peaks, never moving off the peak no matter how hard the wind blows. They form when wind ascending a mountain flank reaches the condensation level at or just above the mountain summit. This cloud may resemble a cumulus cloud if it continues growing ver-

tically from the summit. Often, however, a cap cloud forms a small, stratiform cloud with soft edges around the peak.

Interestingly, cap clouds appear to remain quite stationary over summits rather than drifting downwind as clouds in the higher airflow above them do. The reason is a combination of the wind trajectory over the summit and the humidity level of the air around it. As the air rises to cross the summit, its moisture content reaches the condensation level. After passing the summit, the air sinks on the other side to an altitude below the condensation level, and thus the cloud evaporates. As that air moves away, a new air parcel follows to form a new cloud at the summit. In effect, the cap cloud we see is not one individual cloud but an ever-changing series of clouds in a constant cycle of birth and dissipation. Cap clouds may remain fixed over a summit all day, only dissipating with the ceasing or shifting of the ascending wind or changes in the condensation level.

If the summit cloud becomes elongated downwind like a flowing flag, we call it a "banner cloud." For banner clouds, the air stream remains above the condensation level much farther downwind than for the cap cloud. A similar cloud is the rotor cloud. Here, as the wind blows across the mountain ridge, it eddies, like rapids in a stream, forming a horizontal roll. Like the banner cloud, it remains stationary off the ridgeline, but with its long axis parallel to the ridge rather than perpendicular to it.

Clouds are the sky's own calligraphy; they spell out its intent in crisp letters of water and ice.

Eric Pinder, *Tying Down the Wind*

LENTICULAR CLOUDS

Perhaps the most spectacular cap clouds are the lenticular clouds. The full name for these clouds is lenticular cumulus, further defined into altocumulus standing lenticularis, stratocumulus standing lenticularis, and cirrocumulus standing lenticularis. They may form at the summit, some altitude above it, or ahead of the summit.

A distinctive shape and strong outline characterize lenticular clouds, often taking a layered pattern like a stack of flapjacks. Their name derives from their shape's resemblance to a lens. They typically appear in stable atmospheric layers above the mountain, which restrain their vertical growth.

Like the cap clouds, lenticular clouds do not move from their location relative to the mountain but are constantly being reborn as new air moves through their position. Lenticular clouds have been mistaken for UFOs because they are often shaped like a flying saucer and because they appear to

A stack of lenticular clouds pictured over Mt. Jefferson in Oregon, with Detroit Lake visible in the foreground. They are among the most recognizable of the cloud formations.

hover. Lenticular clouds can form extraordinary cloudscapes when the clouds are caught by the setting sun or when a "fleet" of lenticular clouds forms over a multi-peaked ridge. Often the edges of these clouds are tinged with pastel color, a phenomenon called "irisation."

ROLL CLOUDS

On occasion, low horizontal clouds known as *arcus* clouds develop in association with a thunderstorm or cold front. There are two distinct types of arcus clouds: the shelf cloud and the rarer roll cloud. The shelf cloud, associated with a thunderstorm gust front, generally remains attached to the cumulonimbus.

LENTICULAR CLOUDS

What they are:

A type of cap cloud that forms over mountain summits in the crossing airflow, taking the shape of a lens, often stacked like pancakes. They appear to remain stationary and are frequently mistaken for UFOs because of their saucer-like shape. Often the edges of these clouds may be tinged with color—a phenomenon called irisation.

What happens:

Lenticular clouds form when wind ascending a mountain flank reaches the condensation level at or near the mountain summit. Lenticular clouds appear to remain stationary. As the air rises to cross the summit, its moisture content reaches the condensation level. After passing the summit, the air sinks on the other side to an altitude below the condensation

level, and thus the cloud evaporates. As that air moves away, a new air parcel follows to form a new cloud at the summit. In effect, the lenticular cloud we see is not one individual cloud but an ever-changing series of clouds in a constant cycle of birth and dissipation.

Where to see them:

Isolated mountain peaks form the most spectacular lenticular clouds, but they form over any high terrain when conditions are right and disturb the wind field above it.

When to see them:

Given the proper conditions, they appear in all seasons. The most spectacular formations are seen around sunset when colored by the setting sunlight.

A schematic diagram showing how lenticular clouds form in airflow crossing a mountain ridge.

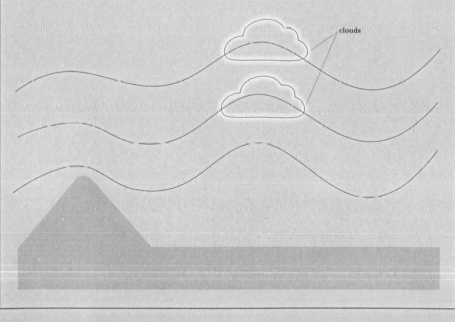

clouds

The thundercloud is a gigantic, if comparatively slow, explosion of moist air, the latent heat of the moist air acting as the fuel.

Sir Napier Shaw, *The Manual of Meteorology*

Roll clouds, in contrast, are not attached to storm clouds and often appear to "roll" across the sky. Long and tubular in form, roll clouds may be a sign of strong downdrafts, such as microbursts, when associated with a thunderstorm. They arise when the cool, sinking air of a thunderstorm's downdraft spreads out across the surface and its leading edge undercuts warmer air entering into the storm as an updraft. Contact with the cooler air condenses the warm air's moisture, creating a cloud. At times, the interaction between the two airflows creates a horizontal rolling eddy—the roll cloud. A similar situation may arise ahead of a vigorously advancing cold front, with the roll cloud standing alone in the sky. A very spectacular and rare roll cloud has been named the Morning Glory (not to be confused with the glory described in Chapter 2).

THE MORNING GLORY

The most spectacular and frequent Morning Glory roll clouds appear in the skies off Northeast Australia's Gulf of Carpenteria, near the Queensland town of Burketown. This cloud got its name from its glorious appearance and the fact that it appears around sunrise and dissipates a few hours later. Although the Morning Glory has appeared at many locations around the world, at Burketown it appears regularly each September and October. The associated wind field produces incredible soaring conditions, making the small Australian town a mecca for glider pilots.

The Morning Glory forms a cloud tube that may stretch over 600 miles (1,000km) and rise to 10,000 feet (3,000m). It advances across the terrain at 35mph (60km/h). Sudden wind squalls, intense wind shears, and a sharp rise in atmospheric pressure accompany the Glory. Ahead of the cloud, strong updrafts lift the air, which then begins to sink near the cloud back, thereby giving a rolling appearance to the complex. Scientific analysis of this wonder describes it as a *soliton*, a solitary wave with a single crest that advances without changing form or speed. Others consider it similar to a tidal bore (see p. 190).

After years of speculation, an acceptable explanation for the formation process has emerged. Morning Glories develop when two air masses of differing temperatures come together. When they meet, the warmer air rises, and the colder air replaces it at lower altitudes. During the night, a cool layer develops at the surface, allowing the colder air mass to advance quickly, producing a "bump" along the cold air mass's boundary. This bump becomes well established by morning, advancing inland against the warm air, like a single wave or tidal bore, and forcing a strong updraft in the warm air. If sufficient moisture is present, it condenses upon rising, generating the visible roll cloud.

The Cape York Peninsula, sitting between the Coral Sea and the Gulf of Carpentaria, allows sea breezes to form on both shores that act up the ideal convergence zone along the western shore, and therefore the Morning Glory frequently occurs there. Similar phenomena have been reported elsewhere in Australia, and over the United States, Atlantic Canada, Britain, Germany, Russia, the Arabian Sea, Mexico, and Uruguay.

NACREOUS AND NOCTILUCENT CLOUDS

This section finishes with two of the rarer and most mysterious members of the cloud family: nacreous clouds and noctilucent clouds. These two cloud types form high in the stratosphere, 10 to 60 miles (16—100km) above the surface and above the troposphere, where all other clouds form.

Nacreous clouds, also known as "mother-of-pearl clouds" and "polar stratospheric clouds," form only in winter when strato spheric temperatures dip to chilling depths, −108°F (−78°C) and lower. As a result, they usually only appear in the polar regions, though they may dip as far south as England on rare occasions. Forming in extremely dry reaches of the atmosphere, they usually contain some water ice and crystalline nitric acid and/or sulfuric acid. Nacreous clouds of pure water ice only form when temperatures fall below −120°F (−85°C). When they do form, nacreous clouds are only visible for a few hours after sunset or before dawn. Their great height allows their illumination by the sun from below the horizon, increasing their brilliance against the dark twilight sky.

Nacreous clouds blaze in the twilight sky with brilliant, vivid iridescent colors like mother-of-pearl. They appear as filmy sheets that shift colors as their crystals move in the fading sunlight. The sheets themselves change shape, stretching and contracting, curling and uncurling slowly in the fading light. In the southern hemisphere, they are

solely Antarctic residents. In winter in the northern hemisphere, nacreous clouds may be seen in Scandinavia, Iceland, Alaska, and Northern Canada, although on rare occasions they may appear in the skies much farther south. A spectacular display in February 1996 surprised Irish, Scottish, and English eyes as far south as London and Cambridge. Much still needs to be learned about the formation of these clouds, but they are implicated in ozone destruction and the creation of the polar ozone holes.

Higher still is the realm of the noctilucent clouds. These mysterious clouds form in the upper stratosphere and lower mesosphere, 45 to 55 miles (70—90km) above the Earth's surface. Many believe that small ice-coated particles account for their composition, though others favor pure dust. They are extremely tenuous; so transparent that they reflect only one thousandth of the incident sunlight they intercept. Therefore noctilucent clouds only become visible when illuminated from below against a dark background. Like nacreous clouds, they only appear in the western sky about 30 minutes after sunset when the Sun lies from 6 to 16 degrees below the horizon. The clouds themselves typically extend from 15 to 20 degrees above the horizon, along the edge of the twilight arch.

The annual observational peak occurs several weeks either side of the summer solstice, from mid-May to mid-August in the northern hemisphere. The most favorable latitudes for sightings lie between 50 and 65 degrees. Farther north, the midnight sun or strong midnight twilight leaves the sky too bright. Farther south, twilight is too short. In the southern hemisphere, ground-based observations are infrequent because this hemisphere has little land area within the favored band.

Precipitation

Precipitation gives us some of the most beautiful of the
atmospheric wonders, from the delicacy of snow crystals to the
dazzling splendor of glaze ice clinging to trees. But it is also a
wonder of efficiency in moving heat around the planet, rivaling
the winds for prominence in the atmospheric heat engine.
Without rain and snow, life on Earth would be quite different,
perhaps only able to cling tenuously to the land
along the coastal margins.

In the previous section, we discussed the processes that form clouds. Some clouds will continue their evolution and produce precipitation. Cloud droplets fall very slowly (0.5 inches/1.2cm per second), which is not what we experience in rainfall, where the average fall rate hits around 15 feet (4.5m) per second. A small raindrop—a drizzle drop with a diameter around 0.05 inch (1.2mm)— contains the equivalent of a million average-sized cloud droplets. In order to begin meaningful precipitation, however, droplets must grow much larger, to a typical raindrop size with diameters of around 0.12 inch (3mm).

THE GROWTH OF RAINDROPS

So that cloud droplets can grow, they must remain in the moist cloud environment, which in turn requires a continuation of the lifting process. The updrafts that initially formed the cloud warm as water vapor turns to liquid or solid states, conversions that involve the loss of latent heat. With this added heat, the ascending parcel warms relative to the surrounding air, increasing its buoyancy, and therefore it continues ascending. We can actually see this rise in cumulus clouds as they boil up to great vertical heights.

Cloud droplets can grow larger in three ways. The first is by the continued condensation of water vapor onto cloud droplets, but compared to the initial droplet-forming condensation, continued growth by condensation proceeds very slowly after a certain size is reached.

A thunderstorm rolls across wheatfields in Alberta, Canada, dropping localized precipitation under its cumulonimbus. The dark thunder cloud, towering high on the horizon, is one of the most formidable sights in the natural world. A dramatic shaft of rain is visible cascading down from the heart of the cloud.

Let the rain kiss you. Let the rain beat upon your head with silver liquid drops. Let the rain sing you a lullaby.

Langston Hughes, "April Rain Song"

Growth by collision and coalescence of cloud droplets, followed by the continued collision of drops with cloud droplets and other raindrops, moves at a much quicker pace. Turbulent air currents within the clouds initiate the first collisions between droplets, producing a drop, which further collides with other droplets, and so on. As drops grow, their fall velocity also increases. Eventually, they become too heavy for the updraft to support them and they begin their descent. While falling, drops may collide with slower-falling drops or those that continue to ascend. Larger drops may still ascend in the updraft but at a slower rate relative to the myriad of smaller drops around them, so the smaller, faster-rising drops overtake the larger ones, "rear-ending" them and coalescing. When drops are too large, however, their collection efficiency for the smaller droplets decreases—small droplets simply bounce off or flow around the larger drops uncollected. Consequently, drop growth slows. The best drop growth occurs in clouds with strong, large updrafts because the drops remain in the cloud longer and therefore have many more collision opportunities.

It may seem odd, but the best conversion of cloud water to precipitation occurs when the cloud contains ice crystals. As a drop, liquid water must be cooled well below 32°F (0°C) before freezing. In fact, a pure cloud droplet may not freeze until its temperature drops to −40°F (−40°C). Therefore, through-out the cloud, ice crystals and water drops co-exist. When ice crystals and supercooled (at a temperature below 32°F/0°C) droplets crowd together, water molecules move from droplet to crystal, increasing the ice crystal's size at the expense of the droplet. When ice crystals grow at temperatures around 14°F (−10°C), they develop arms and branches and become snow crystals. Such crystals not only grow efficiently at the expense of water droplets, but they also easily stick together, congregating in large aggregates called snowflakes.

Eventually, the updraft ceases or the drops (or snowflakes) reach sizes that can no longer be upheld by the updrafts, and they begin their descent to the surface. If the air that they traverse is below freezing, the precipitation falls as snow. If the air layer is above freezing and sufficiently thick, snowflakes melt and the precipitation becomes rain. Most rain falls as drops in the size range of 0.008–0.2 inch (0.2–5mm). Snow crystals are typically 0.02 to 0.2 inch (0.5–5mm) in breadth whereas snowflakes average about 0.4 inch (1cm) across.

Early morning dewdrops are pictured here on long grass. Mist, produced as the heat of the Sun's rays warms the ground, is also visible in the background.

PRECIPITATION TYPES

Most precipitation falls as rain, but not all. Outside the tropics, significant precipitation falls as snow or, less frequently, another form of frozen precipitation. If precipitation is intermittent—stopping and starting suddenly—it is called a shower. Showers are short-lived, rarely lasting more than half an hour, and generally fall from cumulus-type clouds, but they can bring large accumulations. Precipitation events vary both in intensity and in the size of the raindrop or snowflake.

RAINFALL

The lightest rain event, drizzle, accumulates very slowly, at a rate of around 0.02 inches (0.5mm) of water per hour. One factor in drizzle's lightness is its small drop size, typically less than 0.05 inches (1.2mm) in diameter. For comparison, cloud droplets are about 100 times smaller, while the average raindrop is over twice as large. Drizzle normally falls from low nimbostratus clouds as a continual, soft rainfall. Light rainfalls have precipitation rates below 0.1 inch (2.4mm) per hour, while moderate rainfalls drop between 0.1 and 0.3 inches (2.4–7.2mm) per hour. Heavy rainfalls descend at between 0.3 and 2 inches (7.2–50mm) per hour, and extreme rainfalls exceed 2 inches (50mm) per hour. While drizzle has a rather uniform drop-size distribution, harder rains generally have a distinct drop-size distribution, with very few small and very few large drops. The median drop size tends to increase with rainfall rate.

SNOWFALL

Snowfall has a range of intensities as well, expressed in inches or centimeters of snow per hour. Reported observations of snowfall intensity, however, are based on visibility reductions during the snowfall. For example, a heavy snow, accumulating two inches (5cm) or more per hour, reduces visibility to less than a quarter mile (400m).

The ultimate snowstorm is the blizzard. The U.S. National Weather Service defines a true blizzard as having wind speeds exceeding 35mph (56km/h) and falling or blowing snow reducing visibility to under 1,300 feet (400m). In addition, these conditions must last at least three hours. The British Meteorology Office defines a blizzard as a moderate or heavy snowstorm with a mean wind speed of 30mph (48km/h) and visibility below 650 feet (200m). Canadian criteria require wind speeds above 25mph (40km/h) and visibility below 0.625 miles (1km) for at least 3 hours, but each province has slightly different time and temperature criteria.

. . . the snow patches are not rose-colored but a very dark purple like a grape, and thus they are all degrees from pure white to black . . . the snow patches are a most beautiful crystalline purple, like the petals of some flowers or as if tinged with cranberry juice.

Henry David Thoreau, *Journals*

OTHER FROZEN PRECIPITATION

A small percentage of precipitation falls as frozen forms distinct from snow: hail; sleet or ice pellets; graupel or soft hail; snow grains; and freezing rain. All but hail begin as ice or snow high in the clouds, but as they descend they traverse a layer of air with above-freezing temperatures and totally or partially melt, depending on the layer's depth and its departure from freezing. If the precipitation does not enter a subfreezing air layer again before reaching the surface, it falls as rain or a mixture of rain and snow. However, if it passes through a subfreezing air layer before hitting the surface, drops may refreeze, partially refreeze, or supercool. If caught in updrafts and downdrafts, making multiple passes through warm and cold layers, they may form several distinct ice layers, like a hailstone.

Ice pellets, or ice grains, are hard and either clear or opaque ice particles, spherical or irregular in shape with diameters of 0.04–0.16 inch (1–4mm). They form through the refreezing of raindrops or partially melted snowflakes

falling through a cold air layer near the surface. Often called "sleet" in the United States, they bounce when hitting hard surfaces.

Graupel, snow pellets, or soft hail is soft, opaque ice, irregular in shape and often studded with oblong or branched protrusions. Graupel splatters on impact. With diameters of 0.04–0.28 inch (1–7mm), graupel frequently forms in strong updrafts, such as in lake-effect snows when supercooled water coats a snowflake with ice, or when a supercooled drop develops an outer ice coating without freezing through. Because descending convective currents may push graupel quickly downward, it can fall with surface temperatures significantly above freezing.

A thick snow cover whitens a coniferous forest in the Brdy hills, Czech Republic. Snow crystals vary in shape according to air pressure and the relative humidity within the cloud.

[Snowflakes] are about a tenth of an inch in diameter, perfect little wheels with six spokes without a tire, or rather with six perfect little leaflets, fernlike, with a distinct straight and slender midrib, raying from the center.

Henry David Thoreau, *Journals*

Snow grains are tiny, white, opaque particles—the solid equivalent of drizzle—either flat or elongated, and less than 0.04 inch (1mm) in length. Smaller than snow pellets, they generally fall from stratus clouds rather than convective clouds.

"Sleet" is a term for cold-weather precipitation with several different definitions. In the United States, sleet technically refers to ice pellets but sometimes erroneously describes freezing rain. In Great Britain, sleet refers to mixed rain and snow.

Rime and glaze, often called "freezing rain," are supercooled drops falling as a liquid that freeze on contact with cold surfaces. Temperature and fall rates determine which form a freezing rain becomes. Glaze is clear-ice freezing rain. The amorphous, dense structure of glaze clings tenaciously to any surface on which it forms, causing extensive damage. If freezing-rain ice appears milky, it is called rime. In contrast to glaze, rime ice is less dense and clings more loosely, therefore generally inflicting only minor damage.

Frozen precipitation may fall as either showers or continuous precipitation, but some forms have a preferred manner. Ice pellets may fall steadily or in showers. Freezing rain and snow grains usually fall as continuous precipitation, whereas graupel generally falls in showers.

HAIL AND HAILSTONES

Hail, another form of frozen precipitation, differs from the others in that it occurs predominantly in the warm seasons when vigorous thunderstorms rage. Since hail requires strong updrafts, cold-air regions, sufficient ice nuclei, and supercooled water, squall-line thunderstorms and supercell thunderstorms produce the most hail. Supercells have the greatest potential for large, damaging hail since they live longer than regular thunderstorms and reach to the high, coldest levels of the troposphere.

Hail formation begins on seeds of small frozen raindrops or graupel. In order for the hail embryos to grow, they must remain in cloud regions rich in supercooled water for some time, growing by accretion of new layers of ice—the longer the residency, the larger the hailstone's potential size. Because gravity pulls them earthward, there must be a countering force to keep infant hailstones aloft, and this is supplied by the strong updrafts within the great cumulonimbus towers of a thunderstorm. For the smallest hailstone to form, updrafts must rise at speeds exceeding 24mph (40km/h). Larger stones 1.75 inches (4.5cm) in diameter—golf-ball sized—require updrafts above 55mph (90km/h). Softball-sized hailstones, 4.25 inches (10cm) across, need updrafts of over 100mph (160km/h).

Fifty years ago, the widely held theory of hail formation visualized growing hail riding a roller-coaster of thunderstorm updrafts and downdrafts before falling earthward. We now believe this is but one possible mechanism. Hail need not ride the cumulonimbus coaster, but may grow by slowly falling through a layer rich in supercooled water. Eventually, a hailstone grows too big for the updrafts, or gets caught in a downdraft, and plunges to earth. Not all hailstones survive the journey, however. Often less than half the hailstones formed within the cumulonimbus nursery ever touch ground. Some melt as they traverse the warm air below the storm's freezing level, others break up into smaller chunks that melt before reaching the surface.

The U.S. National Weather Service describes hailstone diameters with a series of descriptor terms ranging from "pea-sized" to "softball-sized." The British Meteorological Office uses a slightly different set of terms from "pea-sized" to "coconut-sized." Larger hailstones than softball- or coconut-sized are possible, although they usually result from the joining of several large stones during descent.

The largest hailstone (by dimensions) ever recovered fell in Aurora, Nebraska, on June 22, 2003, with a record 7-inch (18cm) diameter and a circumference of 18.75 inches (47.6cm). It exceeded the previous record hailstone that fell in Coffeyville, Kansas, on September 3, 1970. That stone weighed 1.67 pounds (0.75kg), and spanned 5.67 inches (14.4cm). The official world's heaviest hailstone fell in the Gopalganj district, Bangladesh, on April 14, 1986 and weighed 2.25 pounds (1kg).

SUN SHOWERS

Showers are defined as precipitation falling from convective clouds, characterized by the suddenness with which they start and stop, and rapid changes of fall. A shower may fall from a single cumulonimbus cloud in an otherwise clear sky. Showers may also fall from convective cells scattered like players on a football field across skies that are not completely overcast, so that sufficient clear sky between the clouds allows the sun to shine directly through. In fact, this is usually the situation when a rainbow appears, since it requires sunlight to interact with raindrops. This situation is often called a "sun shower."

Sun showers can produce an interesting situation wherein it actually gets brighter during the rain than before or after it. During a sun shower, the many rapidly falling raindrops reflect light rays off their surface. Sunlight entering the downpour therefore undergoes multiple reflections, causing light to reach our eyes from all directions, increasing the overall brightness. In effect, we are caught among a large number of little falling mirrors.

COLORED RAIN AND SNOW

Eric Burden and the Animals sang of "Colored Rain," while Prince just saw "Purple Rain." Frank Zappa warned us, "Don't Eat the Yellow Snow." Colored rains and snows, however, are not solely imagined by songwriters and poets but are real, naturally occurring phenomena.

In purest forms, snow and rain are nearly transparent but take a blue tinge when large quantities cover the ground. While falling, however, they may collect dust, sand, or pollen grains from the surrounding air (a process known as "scavenging") in sufficient

We want to listen to the wind shout and watch snowflakes the size of bottlecaps streak across the window.

Jerry Dennis, *The River Home: An Angler's Explorations*, 2000

quantity to give the resulting precipitation a distinct color.

Falling rain normally appears white as it reflects the sun's light. When on the surface, its transparency allows a puddle to take on the underlying color or its reflectivity gives the puddle the color of the sky or surrounding terrain. But at times, the rain itself takes on a color. Most colored rains originate from surface materials that have been suspended in the air by a dust- or sandstorm and are therefore hued with earth tones. Storms "raining blood" usually occur when rain scavenges red dust or sand blowing off arid areas such as the Australian desert. In ancient times, such rains were thought to foretell death or the wrath of the gods, but in reality these rains are beneficial, cleansing the air of impurities. Rain that scavenges severely polluted air can take the color of the polluting particles.

Biological elements may also tinge rain, the most striking being yellow rains colored by pollen "storms" when large forests of pine, birch, or spruce trees release their windborne yellow pollen in the spring. During the summer of 2001, colored rains fell on the Indian state of Kerala. A government study found the mostly red coloration came from the spores of a lichen-forming alga belonging to the genus *Trentepohlia*.

Snowfall will also scavenge colored materials from the air as it falls, particularly when associated with dust storms. The coloration may be seen during snowfall, but most snow coloration becomes apparent after it accumulates on the ground. The U.S. State of Colorado frequently receives snows colored by dust picked up from the Great American Desert. For example, in February 2006, Coloradans reported newly fallen snow as chocolate-brown hued.

Once accumulated on the ground, even pure snow takes on a distinct hue. Ground snow initially appears white because it uniformly reflects the full color spectrum of sunlight from its surface. Of course, if the sunlight or skylight has a dominant color, such as around twilight, its reflection off snow will tint the snow's appearance to its hue. If the light penetrates the snow, the snow cover absorbs blue light slightly less than other wavelengths, thus giving it a bluish tint.

Ground snow also takes the color of any impurities, including pollutants, that lie on or just beneath its surface. Algae often colonize snow banks, giving them their color. One alga, *Chlamydomonas nivalis*, produces red and green snow that has been called "watermelon snow." These colonies commonly inhabit the alpine and polar regions. Other algae may tint snow reddy-pink, yellow, brown, or a deeper blue. Windborne tree pollen, spores, and dusts may also color snow when they deposit on the surface. Pine, birch, and spruce can produce a yellow snow somewhat different from that alluded to by Zappa.

Finally, snow-covered mountains may glow with the twilight colors as that light reflects off their snowfields. Particularly wonderful are the "purple mountain's majesty," seen when summits catch the lavender hue of the upper twilight arch, an effect termed the "alpenglow."

 Frost feathers, formed overnight on a door window, are backlit by the morning sunlight. Such structures are often seeded by dirt particles on the glass, producing delicate feather patterns.

SNOWFLAKES

"Snowflake" technically refers to an assemblage of individual snow crystals. Snowflakes typically appear when near-surface air temperatures hover around the freezing point. At such temperatures, snow crystals are more "sticky," and those that collide will adhere to one another better than at colder temperatures. At very cold air temperatures, they do not stick together at all, and therefore bitter-temperature snows drop mostly snow crystals.

While snow crystals usually measure 0.02 to 0.2 inch (0.5–5mm) across, snowflakes are typically bigger, having diameters about 0.40 inch (1cm) on average, with most large flakes being 0.8 to 1.6 inches (2–4cm) across. Exceptionally large snowflakes aggregate thousands of individual crystals. For snowflakes to exceed 2 inches (5cm) across, however, perfect conditions must prevail. Given optimum temperatures for stickiness, large flakes only survive under light winds; otherwise, they break apart as they fall. The largest reported snowflake measured 15 inches (38cm) across, falling at Fort Keough, Montana on January 28, 1887. Another giant—8 by 12 inches (20cm by 30cm)—descended in Bratsk, Siberia in 1971. Two decades earlier, residents of the English town of Berkhamsted reported snowflakes the size of saucers, almost 5 inches (13cm) across, descending from the sky.

SNOW CRYSTALS

We think of snow as flat, six-sided, frilly stars resembling frozen lace. Snow crystal structures, however, take on many forms. Besides six-armed stars, shapes range from flat hexagonal plates to three-dimensional, hexagonal pillars with plates attached to each end—popularly called "cufflink crystals." Other snow crystals resemble Doric columns, six-petalled flowers, traffic signs, and dinner plates. Normally, such perfectly formed crystals do not survive the fall to earth. Many break off arms as they are buffeted by winds within a cloud, and others fuse together, producing irregular and asymmetrical shapes.

For centuries, scientists puzzled over why snowflakes had six sides, even drawing the attention of famed astronomer Johannes Kepler away from the heavens to look at snowflakes. We now understand that snowflakes' shapes begin at the molecular level through the arrangement of the atoms composing the water molecule. Each molecule contains two atoms of hydrogen joined to one oxygen atom—H_2O. Oxygen takes a more powerful hold on the electrons the two elements share in their water-making bond. As a result, the oxygen side of the water molecule becomes slightly more negatively charged while the hydrogen side has a slightly more positive charge. Since opposite charges attract, the positive hydrogen atoms of one water molecule are attracted to the oxygen atom of a neighboring molecule. In the liquid and vapor phases, water molecules jostle and dance around their neighbors, forming only occasional weak links between molecules.

However, when a water mass cools toward the freezing point, the attraction between individual water molecules becomes stronger. Upon freezing, everything seizes up. Now, each water molecule is surrounded by four others to which it bonds. The oxygen atoms are arranged hexagonally in layers. When additional molecules combine on a growing crystal, they need less energy to adhere to the crystal sides than to add a layer on the crystal face. Therefore the side faces grow more quickly, forming the well-known six-fold symmetry of the crystal lattice structure.

Ice crystals are sensitive to the conditions under which they form, particularly the air temperature and the excess relative humidity within the clouds. These factors affect not only how crystals grow but also their eventual shape: the colder the temperature, the sharper an ice crystal's tips. At warmer temperatures, ice crystals grow slowly and smoothly, resulting in less intricate shapes—needles and plates. Thin, hexagonal plate crystals form in temperatures from 25° to 32°F (−4° to 0°C). In slightly colder air (21° to 25°F/−6° to −4°C) needle shapes form. Long, hollow, hexagonal columns appear from 14° to 21°F (−10° to −6°C) while flower-like plates arise at temperatures from 10° to 14°F (−12° to −10°C). Six-pointed stars, dendrites, form in temperatures from 3 to 10°F (−16° to −12°C). Generally, dendrites form in high-altitude clouds; needles or flat six-sided crystals originate in middle-height clouds, and a wide variety of shapes grow in low clouds.

Hast thou entered into the treasures of the snow?

Job 38:22

S N O W F L A K E S

What they are:

Aggregations of small snow ice crystals. Snowflakes are typically bigger, with breadths of 0.4 inches (1cm) on average; large flakes measure 0.8 to 1.6 inches (2–4cm) across. Exceptionally large snowflakes aggregate thousands of individual crystals. Snow crystal structures take many forms. Besides six-armed stars, shapes range from flat hexagonal plates to three-dimensional, hexagonal pillars capped with hexagonal plates. Other snow crystals resemble Doric columns, six-petalled flowers, traffic signs, and dinner plates.

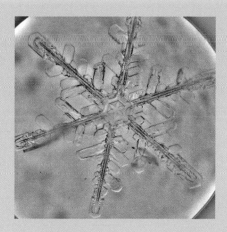

What happens:

A snow crystal's formation begins at the molecular level. Each water molecule contains two hydrogen atoms and one of oxygen. Since oxygen takes a more powerful hold on shared electrons, the oxygen side of the molecule has a slight negative charge. Since opposite charges attract, the oxygen side of one water molecule attracts the positive hydrogen side of a neighboring molecule. On freezing, water molecules arrange themselves hexagonally in layers. Ice-crystal growth is sensitive to the air temperature. At warm temperatures, ice crystals grow as thin, hexagonal plate crystals. As the environmental temperature decreases, needle shapes form next, then hollow hexagonal columns and flower-like plates. Six-pointed dendrites grow at the colder temperatures.

Where to see them:

In the polar half of both hemispheres, occasionally equatorward to about 35 degrees latitude, and at all latitudes at elevation.

When to see them:

Generally during the winter season, but snow crystals may appear at higher latitudes and elevations at any time of the year.

Examples of snow crystals, as photographed by Wilson Bentley (1865–1931).

Humidity-Based Phenomena

Moisture in the air need not all fall out as precipitation. Some of it is deposited on the surface, providing dazzling landscapes of dew and frost. Fog, in contrast, encloses our vision in ambiguity. Each of these humidity wonders has served poets, songwriters, and novelists well, however. They have also filled our mythologies with stories of gods and spirits.

Earlier, we saw how condensation produces clouds and precipitation high in the atmosphere. Condensation, however, does not solely occur in the air. It frequently happens on the ground and on low-lying surfaces, or just above. When on a surface, condensation is called dew or frost; when in the air close to the ground, it is known as fog. Humidity-based phenomena have produced many weather proverbs, such as "When dew is on the grass, no rain will come to pass" and "A summer fog for fair, a winter fog for rain." Frost naturally brings to mind that giant of weather folk heroes, Jack Frost. Thick fog reminds us of the London of Charles Dickens' novels.

What Is Humidity?

Before looking in depth at these entities, a few words on humidity are in order. The discussion is all relative—relative humid-ity, that is. Many use the term "humidity" indiscriminately when they talk about the amount of water vapor in the air. Its defini-tion actually has two major subcategories: absolute humidity and relative humidity. Meteorologists define the former as the "mass of water vapor contained in a volume of air," while the latter denotes "the dimen-sionless ratio, expressed as a percentage, between water's vapor pressure in the air and the saturation vapor pressure for the air temperature." (Vapor pressure is the pres-sure exerted by water vapor in the air and depends on the quantity of vapor present and its temperature.) Under normal surface weather conditions, relative humidity is also the ratio of the absolute humidity in the air and the saturation absolute humidity for the ambient temperature.

The key in the relative humidity defi-nition is saturation. Saturation is reached

Hoar frost on grass stalks in barely visible formations creates an almost fur-like impression to the naked eye. The frost forms slowly—often on cold, clear nights, when there is not much wind—and grows outward in delicate patterns. This photograph was taken in Comox on Vancouver Island, in British Colombia, Canada.

The fog came pouring in at every chink and keyhole,
and was so dense without, that although the court was of
the narrowest, the houses opposite were mere phantoms.
To see the dingy cloud come drooping down, obscuring
everything, one might have thought that Nature lived
hard by, and was brewing on a large scale.

Charles Dickens, *A Christmas Carol*

when the partial pressure of water vapor in the air reaches its saturation value for the existing ambient air temperature, thus giving a relative humidity of 100 percent. When saturated, a mass of air lying over a standing water surface loses as much water vapor through condensation to the water surface as it gains through evaporation, meaning that there is equilibrium in the exchange of water vapor between the air mass and the liquid. With relative humidity below saturation, water will evaporate from the water surface to the air; when above saturation, there will be condensation of vapor onto the water surface. You may hear saturation called the condition where the air contains "all the water vapor that it can hold." Though not technically correct, old metaphors die hard.

Absolute humidity varies only moderately through an air mass, being added mainly by evaporation of precipitation or surface water, subtracted by condensation, or altered by mixing along the air-mass edges. Relative humidity's value, however, varies through the day with the air temperature. It is generally higher in the pre-dawn hours and lowest in mid-afternoon. We call the temperature at which ambient air reaches saturation the "dewpoint temperature"

or simply the dewpoint. Its value solely depends on the absolute humidity of the air. If the dewpoint is below freezing, it is called the "frostpoint." A parcel of air can become saturated by cooling the parcel down to its dewpoint, or by increasing its water-vapor content through evaporation or mixing with more humid air. Both processes are at work continually in the atmosphere, but cooling is usually the process for forming dew and frost and is often the cause of fog.

DEW

The rising sun's first rays illuminate a landscape that shines with a bejeweled light. Visible moisture beads all the surfaces exposed to the air and sky—grass, flowers, leaves, and car windshields—despite the fact no rain fell overnight. Contrary to the oft-used expression, "dew fell overnight," dew does not fall; it somewhat magically appears, like a fairy-tale apparition. To the trained weather eye, morning dew reveals that the predawn hours were calm and clear, allowing surfaces to cool to the dewpoint.

To produce such beauty, the formation of dew requires the ambient temperature of the surface on which the dew forms, such as vegetation or car exteriors, to drop below the dewpoint temperature. At saturation, the liquid and vapor states of water reach equilibrium, neither gaining nor losing water on a surface. But below the dewpoint, the equilibrium becomes tilted toward liquid gain on a surface as more vapor condenses than liquid escapes by evaporation.

When the surface air layer cools below the dewpoint, most surfaces collect dew. However, if the air temperature is just above the dewpoint, dew deposits on some surfaces while not on others. The reason is that not all surfaces cool at the same rate. Some, such as metal and glass, cool faster than other surfaces, such as wood or vegetation. For this reason, dew forms on automobile bodies and windshields first, and sometimes only there.

On a flat surface, water congregates into large circular dewdrops. However, on non-flat surfaces, its chemical and physical properties work with gravity to pull some rounded dewdrops into elongated shapes that may resemble teardrops as they cling to the edges, such as on leaves and twigs.

The weather saying "When dew is on the grass, no rain will come to pass" is a very old one, and it often accurately forecasts that day's weather. Dew usually forms during clear nocturnal skies commonly found within high-pressure (anticyclone) cells. Anticyclones have sinking air within their centers, which opposes the formation of precipitation-bearing clouds during the day. A similar, but opposite, saying links a lack of morning dew with coming precipitation. In such situations, low nocturnal cloud cover prevents the temperature from dropping to the dewpoint and may be a precursor to precipitation.

Dewdrops cling to plant stalks like teardrops, the combined result of gravity, water surface tension, and nonabsorbent plant surfaces.

THE BRUSH OF JACK

Frost, the cold cousin of dew, usually appears during clear, cold nights with light winds when the air temperature near the ground falls below the frostpoint—the temperature at or below 32°F (0°C) to which air must be cooled to achieve saturation with respect to ice. When the general surface air is warmer than freezing, frost may nonetheless develop on colder surfaces, such as glass or metal, which lose heat more rapidly through radiative cooling. This is why car windshields can frost over while no frost forms on nearby vegetation that cools more slowly.

Such surfaces, we say, have been visited by the legendary weather artist Jack Frost. Jack is a benevolent character compared to other mythical frost beings. Born Jokul ("icicle") Frosti ("frost"), the son of the Norse god of wind Kari, he immigrated to England with the Norse, becoming Jack Frost, an elf-like being who colors the leaves of trees in the fall and paints frost patterns on windows. Other frost beings include Germany's Old Mother Frost. When she shakes the white feathers from her bed, they fall as snow. The Frost Woman and Frost Man, important weather deities in Finland and northern Russia, control blizzards and other wintry elements. Elsewhere in Russia, Father

Frost is a mighty blacksmith who forges great chains of ice to bind water to the earth each winter. In Japanese folklore, the Frost Man was the roguish brother of the Mist Man. Australian aborigines attribute frost to icicles thrown down to earth from seven sisters whose bodies sparkle with ice. These frosty sisters could not live with men on earth, so they sought a home in the heavens, each a star of the constellation Pleiades.

Surface frost formation may be sporadic across a region, particularly in hilly or mountainous terrain. Since cold air is denser than warm air, it flows like water downhill, pooling in low areas or hollows. Low-lying areas are more susceptible to frequent frost formation than higher elevations, even those only a few feet higher. Called "frost pockets" or "frost hollows," they are poor places for planting gardens, crops, or orchards, or for locating homes.

Frost can form in either of two ways. At temperatures from 32°F down to 0°F (0°C down to −18°C), frost forms initially when water vapor condenses as a supercooled liquid and then quickly freezes. Once the first ice crystals have established themselves, further frost accumulation may proceed on these crystals by vapor deposition directly to ice. On certain surfaces, frost can be depos-

I see, when I bend close, how each leaflet of a climbing rose is bordered with frost, the autumn counterpart of the dewdrops of summer dawns. The feathery leaves of yarrow are thick with silver rime and dry thistle heads rise like goblets plated with silver catching the sun.

Edwin Way Teale

ited directly at these temperatures without these initial ice crystals. In such cases, a characteristic of the surface or impurities on it can seed the process by mimicking the structure of an ice crystal, providing the growth site. This can be a surface flaw (such as a scratch on glass), a chemical impurity, or particles lying on the surface. At colder temperatures, ice deposits directly on the surface from the vapor.

While frost provides great visual beauty across a landscape, it can best be appreciated close-up. The delicate crystal frost patterns make even the most common surfaces a treasured vision. Take a few minutes to study a variety of frosted objects through a magnifying glass. The vision will open a new world of intricate natural wonder.

HOAR AND RIME

There are two frost forms—rime and hoar—that deposit over larger areas, clinging to surfaces and changing landscapes into winter wonderlands. They differ in their method of formation and appearance.

Rime frost appears under rapid frost formation, usually in conditions of high atmospheric water content (vapor and/or liquid) and moderate wind speeds. Rime commonly forms when clouds or fogs of supercooled or near-freezing droplets contact subfreezing surfaces. Often this occurs when moisture-laden clouds move inland from the ocean and over cold coastal mountains. In some cases, depositions can accumulate several feet (a meter or more) thick, growing in the

Delicate hoar frost forms on pine needles during a cold, foggy night in Valemount, British Columbia, in Canada.

direction of the prevailing wind. On close inspection, rime has a grainy appearance like block salt, forming thick spikes or coatings without recognizable crystal structure. Deposited rime frost is opaque and dense, and harder than hoar frost.

Hoar frost covers the branches of leafless deciduous trees in this idyllic winter scene. The presence of air bubbles gives the frost its white appearance.

Hoar frost forms from the slow deposition of water vapor directly to ice. By accumulating slowly, hoar frost forms interlocking crystals growing outward from the surface. Hoar frost generally develops a feather, fern, or flower pattern growing out of the initial seed. Hoar-frost formation requires light winds, which are often present during clear, cold nights, and it is therefore the most common type of frost seen by those dwelling in non-mountainous areas. When we speak of frost on leaves and other surfaces, we are usually referring to hoar frost rather than rime.

Hoar frost shows a well-defined, delicate, crystalline structure growing in steps or layers. Small air bubbles trapped in the ice reduce its transparency, causing the crystal's white color. Smooth crystal faces make hoar glitter in the sunlight, particularly at low, early-morning, sun angles, in sharp contrast to rime frost, which shows a dull, matte finish. Hoar-frost crystals usually stay small, but if left undisturbed under the proper conditions they may grow to large frost ferns and frost flowers. In the quiet conditions found in an ice cave or ice-filled sinkhole, for example, they may rival true ferns and flowers in size.

Fog

The obscuring nature of fog has made it a metaphor for the hidden, the unknown, or the unclear. Its imagery fills literature, song, and poetry, one of the best-known being American poet Carl Sandburg's six-line poem "Fog." Reportedly, the last words of another American poet, Emily Dickinson, were: "The fog is rising."

Fog is technically defined as an obscurity in the atmosphere's surface layer caused by a suspension of water droplets, with or without smoke particles present, and defined by international agreement as being associated with visibility less than 0.62 miles (1km). Often, however, fog patches dot the land-

It was a foggy day in London, and the fog was heavy and dark. Animate London, with smarting eyes and irritated lungs, was blinking, wheezing, and choking; inanimate London was a sooty spectre, divided in purpose between being visible and invisible, and so being wholly neither.

Charles Dickens, *Our Mutual Friend*

Fog is blinding, stifling, somehow. It traps me, makes me feel a prisoner. It has no life, no movement, no beauty. Fog is merely wetness that hasn't settled on the grass. To me fog settles and mist rises.

Hal Borland, *Book of Days*, 1976

scape or hang over one small area. This, too, is fog even though it does not fit the visibility requirement for weather observations. Fog cloud droplets are very small and therefore fall to the ground very, very slowly, appearing to hang suspended in the air. Warming the air or mixing it with drier air can return the liquid water to the vapor state, so fogs may dissipate quickly with the morning sun or an increase in wind speed.

Although we may not recognize it as such, fog is actually a cloud formed at, or descended to, ground level—those patchy fogs that fill low terrain are cloud fragments. Like clouds, fogs emerge when an air mass saturates with water, the vapor condensing to form liquid droplets. The most common processes that bring an air mass to its condensation temperature are:

• The ascent and resultant cooling of an air parcel

• Radiative heat loss from the air parcel

• Mixing an unsaturated air parcel with a cooler one.

We define types of fog by the process through which they reach saturation. Radiation fog and frontal fog, perhaps the most common types, usually arise in areas away from large water bodies. Over and near water, we find fogs classed as "advection"

fogs or sea fogs. In hilly or mountainous terrain, one might experience "upslope" or valley fogs. Depending on how a particular fog forms, it can rise, fall, roll in, or just envelop.

Radiation fog forms when air near the surface cools to its saturation temperature by radiational cooling, usually at night. A wet surface—moist soil or pools of standing water—significantly increases the likelihood of radiation fog formation, so the potential for radiation fog is higher after a rainfall, particularly if followed by a cold front that clears the skies and dramatically lowers the air temperature.

Radiation-fog layers vary in depth and horizontal extent from shallow, scattered patches to a regional blanket as much as 1,000 feet (300m) in depth. Visibility within the fog drops as the density of water droplets in the air increases. Continued cooling can make the fog layer optically denser, often reducing visibility to the proverbial "can't see your hand in front of your face" thickness. Often, patchy radiation fog forms in the countryside in low areas of the terrain, particularly during fall when longer clear nights are conducive to deep cooling. These become preferred fog locations because cold air, being denser than warmer air, flows like water toward the lowest points in the terrain.

Radiation fog dissipates when the air mass is reheated by the sun, or when the wind speed increases, mixing the foggy air with warmer or drier air around or above it. When solar radiation penetrates the fog to heat the underlying surface, it evaporates the fog from below, a process that can be rapid in summer and excruciatingly slow during winter.

Advection fog develops when air moves (advects) over either a cooler or a moister surface, allowing the advancing air mass to reach saturation. Most often, this occurs when moist air moves over cold surfaces such as a large body of cold water, snow, or ice cover: a condensing advection fog. While over the cold surface, the lowest layer of the advecting air mass cools to condensation. Advection fogs often persist, since the weather situation forming them can continue for days.

Evaporation advection fog occurs when a cold air mass crosses a warmer water body so that water vapor evaporates into it, increasing its absolute humidity. Often, these fogs arise when cold air flows off the land and crosses large, relatively warmer waters such as the sea or large lakes. Such advection fogs, called sea fogs, are common in winter. They may also be called "sea smoke" because the fog rises from the surface in moist air plumes that resemble smoke.

Upslope and valley fogs, two special cases of advection fogs, develop in hilly or mountainous terrain. When air ascends over a terrain obstacle, it cools an amount depending on the rise in height. If the air temperature drops below the dewpoint, the resulting condensation will form upslope fog. Valley fogs occur when air from high terrain cools, and, being heavier, sinks into the surrounding valleys. If the air has cooled to the dewpoint, a fog will fill the valley. If the valley is a river valley, evaporation from the river may provide the moisture needed for fog formation.

Frontal fog, also called "precipitation fog," occurs when rain falling from warm air aloft evaporates at or near the surface under light wind conditions. If the moisture content of the air increases until condensation is achieved, fog appears. Such fogs usually form in the vicinity of warm or stationary fronts. Precipitation fog can also arise when hot surfaces are quenched by cool showers. When the rain hits the hot surface, it rapidly evaporates increasing the vapor content to the condensation level.

"PEA-SOUP" FOGS

Natural fog can be very dense at times, reducing visibility to near zero. However, in England after the start of the Industrial Revolution, fogs became even more ominous as smoke and soot—from the burning of soft coal—mixed with natural fogs to make them even denser. Such fogs earned the name "pea-soup fog" or "pea-soupers" because they had the density and dirty yellow-brown color (due to the coal's sulfur content) of pea soup, but certainly not its flavor, for the fog was described as having an acrid taste and unpleasant odor.

The New York Times may have published the first significant reference using this descriptor. An 1871 article described London as a place "where the population are [sic] periodically submerged in a fog of the consistency of pea soup." London and other areas of England had been subjected to such fogs as far back as the 13th century, when King Edward II prohibited the burning of coal while Parliament was in session. The problem increased over Britain with the onset of the Industrial Revolution in the late 18th century.

In about 1807, the dense fog around London became known as "London particular," a term that Charles Dickens used in his 1853 novel *Bleak House*. In fact, many of his novels and stories, including *A Christmas Carol*, featured graphic descriptions of such dense foggy conditions. On December 27, 1813, a pea-souper descended across southern England that lasted nearly a week. Reports say it was so thick that the other side of the street disappeared from view and transportation was slowed to a crawl. A century later, Winston Churchill wrote in *A Traveller in War-Time* (1918): "London had been reeking in a green-yellow fog." Monet's impressionistic painting of the Parliament Buildings in 1901 showed the structure's profile dimly visible in the thick air.

In 1905, Dr. Henry Antoine Des Voeux presented a paper to the Public Health Congress entitled "Fog and Smoke," in which he coined the term *smog*, meaning "smoky fog." The London problem peaked in December 1952, when a Great Smog enclosed the city, killing around 4,000 people over four days, with another 8,000 fatalities attributed to the foul air in the following weeks and months. Thereafter, efforts in Britain and elsewhere to clean the air of sulfur and smoke have virtually eliminated the problem. Today, "smog" refers to photochemical hazes such as those that prevail over large cities.

"Pea-soup" fog was denser than natural fog because the sulfur compounds and soot in the air attracted small water droplets, which combined to produce larger drops that more effectively reduced visibility. Today, very dense fogs may be called "pea-soupers" though they are rarely the original London recipe.

FREEZING FOG AND ICE FOG

Two forms of fog exist that are related to ice: ice fog and freezing fog. Ice fog or Arctic mist forms at very low temperatures as a thin fog composed of small, suspended ice crystals. It rarely forms above $-24°F$ ($-31°C$) unless abundant condensation nuclei are present, and always appears at temperatures below $-47°F$ ($-44°C$), when spontaneous freezing occurs for all drops. Also known as "diamond dust," because the crystals sparkle in the light like gemstones, ice fogs may cause halo phenomena when the sun shines through them.

Areas where natural liquid water or vapor is present at these temperatures frequently develop ice fog. These include very fast-moving streams, open seas, volcano vents, or geysers (see p. 169), and over herds of mammals such as caribou. (Native hunters in Arctic regions often used clouds of ice fog hanging over caribou herds to locate them.) Vapor from combustion, including home heating and vehicle engines, may also provide the necessary water vapor. As a result, ice fogs often envelop polar settlements.

Freezing fog, on the other hand, is fog composed of supercooled water droplets.

The fog comes on little cat feet.

It sits looking over harbor and city on silent haunches

and then moves on.

Carl Sandburg, "Fog"

Generally, liquid water droplets of the size found in clouds and fog remain as liquid down to temperatures of around 14°F (−10°C). Such droplets are supercooled, their temperature being below 32°F (0°C). Under supercooled fog conditions, any contact between the drops and another object results in immediate freezing, leading to rime and hoar frost formation on trees, utility poles and wires, and towers. Conditions for freezing fog development arise when fog moves from over warm water bodies to cold inland surfaces.

Another cause of freezing fog involves the lifting of a cloud up mountainous terrain, where the air temperature falls below freezing through ascent and the droplets lose their heat to the air. If only freezing fog is present, ice accumulation will be small, as most of the deposited ice sublimes back into the air. However, if freezing fog is accompanied by freezing drizzle, enough ice may build up to cause damage.

Big Ben and Parliament are obscured by fog. The "pea-soupers" of the 19th and 20th centuries reduced visibility even further.

Frost Flowers and Ice Ribbons

Looking at a glass pane on a frosty morning reveals that the brush of Jack Frost has left intricate patterns painted there. Elsewhere, however, if we venture forth on cold mornings, we may see that Jack is a sculptor as well, producing delicate three-dimensional frost sculptures on water surfaces and dead plant stems in the woods. These sculptures remind us of flowers, though some have patterns like ferns and ribbons.

Several natural phenomena go by the name of "frost flowers," each having its own intricate beauty. The most common frost flowers, also called "frost feathers," are painted on windowpanes and automobile windshields by Jack Frost. The second, occasionally dubbed a frost flower, arises from hoar-frost growth. A third form—fern-like crystal clusters—blooms on fresh, thin sea ice in polar regions. A final form appears growing out of dead plant stems as extruded ribbons that twist and curl, forming flower-like ice figures.

FROST FLOWERS

Window frost flowers grow on the inside of a building's glass panes when it is exposed to cold air on their outer surfaces and comparatively warm, moist indoor air on their inner surfaces. The inside surface loses its heat through the glass, dropping below the frost point, allowing the moist air to deposit ice crystals on it. On vehicle glass, usually the outer surface cools below the frost point of the air around it. Imperfections in or on the glass, such as scratches or dust particles lying upon it, determine the shape of the ice crystals that form.

Sea-ice frost flowers burst into bloom on delicate surface ice when the air above the sea is still and dry, and much colder than the water below. The key to their growth appears to be the large temperature difference (as much as 36°F/20°C) between the air and water. Because they are so delicate, sea-ice flowers sublime quickly when hit by sunlight. They are usually visible only in the early morning or in shaded areas. Similar frost flowers have been seen on freshwater ice.

ICE RIBBONS

Perhaps the most fascinating ice flowers emerge from plant stems. They have been called frost flowers, ice flowers, and ice ribbons—the latter preferred by Professor Emeritus James R. Carter of Southern Illinois University, who has researched these phenomena more than anyone else. To avoid confusion, hereafter they are called "ice ribbons." Not only are these wonders not true flowers, they are not technically frost either, but hard ice. However, they are usually linked with vegetation, forming out of dead plant stems. In the United States they appear to favor certain species, including white crownbeard (*Verbesina virginica*), yellow ironweed (*Verbesina alternifolia*), frostweed (*Helianthemum canadense*), and wild oregano (*Cunila origanoides*).

Ice ribbons can grow to the size of a hand or oak leaf, with parallel ridges visible in the thin ribbons. Because of their delicacy, they quickly vaporize when temperatures rise above freezing, which is a principal reason for their rarity. Solar heat undoubtedly destroys ice ribbons, and therefore most are found in shaded spots. They emerge when air temperatures sit below freezing but the water within the soil remains unfrozen.

Ice ribbons emerge from the dead plant stems through ruptures between 2 and 4 inches (5–10cm) wide, producing a uniform ribbon shape rather than a cylinder. Perhaps the stem rift tears open when the water within first freezes. Carter speculates that water initially flows from the warmer soil below, then freezes in the upper stem. As freezing in the stem progresses, it squeezes outward like toothpaste, forming the characteristic ribbon-like form. As it emerges, the ribbon may curl and loop as gravity and other forces affect the extrusion. The ice ribbon initially remains plastic—more solid than liquid but not so solid as to be rigid. However, shortly after leaving the stem, the ribbon fully solidifies. Carter believes that the stem physiology of those species from which ice ribbons emerge must be important; otherwise, all woody weed species would sport ribbons.

These photographs are both examples of ice extruded from plant stems and curled into delicate ribbons of fine white ice.

Frost flowers are wildings, outdoor growths created by humidity in the starlight. I find them on new black ice on the river but they also form on ponds and lakes. Pure ice crystals, they look like tufts of snowy feathers.

Hal Borland, *Book of Days*, 1976

CHAPTER FOUR

electrical

phenomena

The distant flashes and low rumbles speak of an approaching thunderstorm. For millennia, storms were believed to be the wrath of vengeful gods, until Benjamin Franklin and others uncovered lightning's somewhat more earthly origins: static electricity.

Fire in the Sky, Fire at Sea

Prior to the 18th century, atmospheric electrical phenomena
were bathed in mystery and superstition, until scientists such
as America's founding father, Benjamin Franklin, provided the
initial understanding of lightning. Over two centuries later,
however, we have yet to unravel all its old mysteries, and we
have added new wonders to the genre with exotic names
such as ELVES, sprites, and trolls.

We live in an electric field although, except for a few instances, we are never aware of it as we go about our daily business. As you read this, thousands of thunderstorms rage across Earth, zapping it with 100 lightning bolts each second, and raining charged drops to the surface. This thunderstorm activity builds an atmospheric electrical potential—a voltage difference of between 200,000 and 500,000 volts between the ionosphere (50 miles/80km above sea level) and the surface. However, without constant regeneration by thunderstorms, this large earth–atmosphere voltage difference would dissipate in only five minutes.

Generally, the planetary surface has a negative charge while the atmosphere above takes a positive one. Down near the planet's surface, that fair-weather electrical potential measures about 30–40 volts per foot (30cm), which means your head sits some 200 volts higher than your feet. When thunderstorms rumble overhead, that potential can increase more than tenfold. Fortunately, the atmosphere, an excellent electrical insulator, usually lacks enough free charge—unattached electrons and positive ions—to produce a dangerous current.

Electrical current flows invisibly from higher potential to lesser potential. However, with sufficiently strong electrical potential, collisions among free electrons, ions, and air molecules excite them into luminosity. If collisions are confined to a small volume, the luminosity becomes visible as a bluish-white glow. Scientifically known as "corona discharges" or "point discharges," they occur on objects, especially pointed ones, when the electrical potential strength exceeds 2,500 volts per inch

A nocturnal thunderstorm blazes as cloud-to-ground lightning flashes earthward, illuminating its underbelly in red light. In dry climates, lightning strikes on trees can be the cause of devastating forest fires. Elsewhere, tall buildings such as skyscrapers can be the site of multiple strikes each year.

(1,000 volts/cm). Under normal circumstances, the atmospheric electrical field reaches such potential strengths only during thundery weather. When thunderstorms become heavily charged, grass blades, twigs, and the horns of cattle may glow at their tips.

Point discharges can form, however, at lower field strengths. The top of an electricity-conducting object rising from the ground, such as a metal tower or mast, has the same potential as its base (which is to say that it is grounded). If the atmospheric electrical field decreases by 5 volts per inch (2 volts/cm)—common in initial thunderstorm stages—the potential at the top of a perfectly conducting object 100 feet (30m) tall would be 6,000 volts higher than the air surrounding it, sufficiently strong to form a point discharge at its top. Electric force also concentrates around a sharp point, which lowers the potential required to form a discharge.

St. Elmo's Fire

The above situation occurs around the points of sailing-ship masts, producing the best-known point discharge: St. Elmo's Fire, also known as *corpusante* and Corpus Santos. The name comes from an Italian derivation of Sant' Ermo (St. Erasmus, c. 300 AD), the patron saint of early Mediterranean sailors. The Fire appears as ghostly dancing flames, sometimes resembling fireworks and occasionally accompanied by hissing sounds. Usually blue or bluish-white in color and with a lifetime of just minutes, the significance of its heatless and non-consuming flame is strengthened by its association with spiritual intervention. Sailors regarded St. Elmo's Fire as a good omen, because it often appeared in the dissipating stages of severe thunderstorms as violent surface winds and seas abated—the answer to the sailor's prayers. When it appeared before a storm or during fair weather, sailors felt that St. Elmo's guiding hand accompanied them.

With the advent of Benjamin Franklin's lightning rod, church spires, and metal weather vanes, St. Elmo's Fire came inland, especially in North America's thundery weather, inspiring tales of ghosts and spirits. Franklin correctly equated the Fire to atmospheric electricity in 1749. He believed his lightning rod could draw the electrical fire "out of the cloud silently . . . and a light would be seen at the point like the sailor's corpuzante." Today, the Fire may glow on aircraft flying through heavily charged skies, appearing along wing tips, propellers, and antennas.

References to St. Elmo's Fire are found in mariners' logbooks, including those of Columbus and Magellan, in the works of Shakespeare, the novels of Herman Melville, and in Charles Darwin's notes from the *Beagle* voyage.

[A] ghostly flame which danced among our sails and later stayed like candlelights to burn brightly from the mast . . . When he appears, there can be no danger.

Christopher Columbus, *2nd Voyage Journal*

Lightning:
Fire Thrown from the Heavens

They flash suddenly out of the sky with the power to explode trees and rock, or to set prairie grass on fire, and then generate ear-shattering blasts. No wonder cultures around the world have attributed lightning bolts to angry or vengeful gods. On average, lightning kills more North Americans than any other storm-related event except flooding. Despite years of study, much about lightning remains a mystery, particularly the elusive ball lightning.

Lightning strikes . . . sometimes out of a blue sky. Its power can split the greatest tree and its voice can split an eardrum. No wonder this natural wonder ranks as one of the most feared weather phenomena and is the subject of myth and legend. Nearly all polytheistic cultures have a lightning god, or a thunder and lightning god, and although we have a general understanding of lightning and its related phenomena, not everything is known.

In the simplest terms, lightning is a large spark of static electricity, similar to the sparks that annoy us at home and work but with much greater power. Since air provides rather good electrical insulation against the movement of electricity through it, "pockets" of differing charge (negative and positive) can develop in the atmosphere, often producing electric potentials measuring millions of volts between a cloud and the ground, between two clouds, or between two areas within a cloud. When this potential "short circuits" and current flows, a lightning bolt is generated.

WHAT CAUSES LIGHTNING?

We still do not completely understand how these "pockets" of differing charge form in the atmosphere. The most accepted theory states that electrons are stripped off precipitation "particles"—raindrops, hail, and snowflakes—by collisions between them while caught in thunderstorm updrafts and downdrafts. When they lose an electron, particles become positively charged and many are swept by storm updrafts to the heights of the thundercloud. Those particles that gained the electron take a negative charge. These are usually heavier particles, such as hailstones, that aren't carried as high. As a result, they form pockets of negative charge in the middle and lower cloud regions. At the same time, a region of positive charge develops on the normally neutrally charged planetary surface because the negatively charged lower cloud repels the negative charges on the surface beneath it while attracting the positive charges. (Similar charge differences occur between clouds or areas within clouds.)

In the month of February about the second watch of the night, there suddenly arose a thick cloud followed by a shower of hail, and the same night the points of the spears belonging to the Fifth Legion seem to take fire.

Julius Caesar, *Commentaries*

As charged regions build within the cloud, so too does the attraction between the pockets of positively and negatively charged particles. Initially, the insulating ability of the air keeps them apart, but eventually this is overcome. At first, a weak, negatively charged streamer of ions moves out of one pocket, along the path of least resistance, toward a positively charged area. This streamer advances in small, discrete steps about 160 feet (50m) long, which take a microsecond to complete. This creates a sinuous, ionized channel about the width of a finger, called a "stepped leader." The voltage potential at the stepped leader's tip can exceed 10 million volts as it links regions of least resistance with an irregular path—much like the non-straight path we would take while running through a forest to avoid the trees.

Eventually, the stepped leader connects with an area of positive charge, and a lightning bolt races back up the leader channel to relieve the voltage pressure, illuminating the jagged leader channel in passing. Several flashes are usually required to relieve all the potential charge. A complete lightning bolt—composed of a series of individual flashes typically about 30 microseconds long—lasts about four tenths of a second. When the stepped leader connection is made from the cloud to the ground, the resulting initial lightning flash actually moves up from the ground to the cloud. Usually we describe cloud–ground flashes as "forked lightning," which has luminous branches off the nearly vertical direct bolt.

If the above process occurs within the cumulonimbus cloud, we call the bolt "intra-cloud lightning," although it too can be forked. These intra-cloud bolts account for about 75 percent of all thunderstorm lightning. Sometimes we call intra-cloud lighting "sheet lightning" because the narrow channel is hidden from direct view, and reflects off the surrounding cloud. We therefore see the broad flash of the reflection illuminating an area of the cloud in a sheet of light rather than a bolt. If lightning's path connects with the ground, it is a "cloud-to-ground" strike. When between clouds, it is "cloud-to-cloud" lightning. Another cloud-to-cloud manifestation called "spider lightning" travels a long horizontal path across a cloud's underside.

Although we usually consider thunderstorms as summertime or warm-season phenomena, they can develop during snow events called "thundersnows." Once considered rare, more observations have been made recently in the United States away from the traditional "home" of the thundersnows, the North American Great Lakes. Most thundersnow events develop during lake-effect snowfalls that form when very cold air crosses the relatively warm Great Lake waters during the late fall and early winter. The large temperature contrast between the lake's surface waters and cold air rushing over them generates strong convection, which explodes into thunderstorms. The strong updrafts push cloud tops and snowflakes to great elevations, allowing for the separation of charge. Since the surface air is very cold, the precipitation formed within the clouds remains frozen as it falls to the surface. The scattering of light and sound waves by the heavy snows diffuses the lightning, creating sheet lighting and muffling the thunder.

Lightning has filled folklore and mythology for millennia, including the lightning gods: Zeus, who threw lightning bolts from atop Mount Olympus, and Thor, whose hammer, Mjollnir, resounded off the anvil-topped clouds, sending sparks in all directions. When the Great Thunderbird of

Four cloud-to-ground lightning bolts illuminate a line of cumulonimbus while another, from cloud to cloud, flashes upward.

Native American mythology beat its wings, its bright feathers flashed across the sky. Often these deities are also those linked to thunder. For example, the Thunderbird's wing beat drummed out as thunder.

Greek mythology tells us that Zeus received his thunderbolt weapons from the Cyclopes, gigantic one-eyed monsters, whom Zeus freed from imprisonment. He used these thunderbolts to assume sovereignty among the Titans. Afterward, the Cyclopes continued serving Zeus as his smiths, forging thunderbolts on Mount Olympus. Three of them—Arges (thunderbolt), Steropes (lightning), and Brontes (thunder)—later ascended to the status of storm gods.

High, pointed structures, such as church steeples, provide favorite targets for lightning. For many years, European church officials tried to ward off lightning strikes by ringing the steeple bells. Not only did this fail to protect the building but it led to the deaths of many bell ringers, for when lightning struck the steeple, the current ran down the wet bell rope and killed the ringer. Protection finally came with the addition of Benjamin Franklin's lightning rod.

Franklin had been studying electricity for many years when in 1749 he drew the conclusion that lightning was indeed electricity. Combining this concept with his other electrical research and strong practical bent, Franklin believed that an iron rod 8 to 10 feet (2.5–3m) long and sharpened to a point at the top might provide protection to structures. He wrote, "May not the knowledge of this power of points be of use to mankind, in preserving houses, churches, ships, etc., from the stroke of lightning, by directing us to fix, on the highest parts of those edifices, upright rods of iron made sharp as a needle . . . Would not these pointed rods probably draw the electrical fire silently out of a cloud before it came nigh enough to strike, and thereby secure us from that most sudden and terrible mischief!" After Franklin had successfully experimented with lightning rods on his own house, he had them installed on the Academy of Philadelphia (later the University of Pennsylvania) and the Pennsylvania State House (to become Independence Hall) in 1752. Even King George III equipped his palace with a lightning rod.

Lightning lore abounds but most is wrong. Some can be dead wrong. Take, for example, the saying "Lighting never strikes the same place twice." In truth, lightning strikes high structures, such as New York City's Empire State Building, Chicago's Willis (formerly Sears) Tower, and Canada's CN Tower, hundreds of times annually, including multiple strikes during a single storm—not counting the multiple strikes hitting within a single bolt. Even people can be hit multiple times. Lightning struck U.S. park ranger Roy Sullivan seven times between 1942 and 1977. Known as the "human lightning rod," he miraculously survived all these strikes. Actually, many struck by lightning survive—from 5 to 30 percent—though often with severe burns. They do not get fried to a crisp as depicted in cartoons, nor are they electrified. Often the difference between life and death is the speed with which the victim receives CPR.

The beliefs that lightning always strikes the tallest object and that lightning only

During those storms the holy body, that is to say St. Elmo, appeared to us many times in light ... on an exceedingly dark night on the maintop where he stayed for about two hours or more for our consolation.

Chronicler of Magellan's voyage

strikes good conductors such as metal are false. Although lightning leaders prefer taller objects and metal ones to relieve the built-up potential, they will take any path available. Lightning can strike the ground, a water surface, or even the sides of a tall structure. Metal is the best conductor, but as any forest-fire fighter knows, trees make excellent targets for lightning strikes.

Among the most dangerous myths is that there is no danger from lightning unless it is raining. In fact, as we shall see, lightning can strike as much as 10 miles (16km) ahead of its spawning thunderstorm—as well as from behind the storm. And while discussing lightning danger, we must mention that one need not be directly struck by lightning to be killed or injured. Two of the largest death tolls from lightning have come on soccer pitches. In Honduras, 17 people died and 35 were injured while in a shelter when lightning struck nearby. In another tragic incident, five people were killed and 100 injured at a game in Malawi in 2001, when the wet field was hit.

According to the U.S. National Weather Service, 73 people die from lightning strikes each year and hundreds more suffer severe injuries. In Australia, the annual average figures are five to ten deaths with well over 100 injuries. Even being indoors does not assure safety from lightning. It is estimated that about one percent of all lightning deaths in the U.S. result from people talking on a corded phone during a thunderstorm. Indoor dangers can also come from computers and other electrical equipment, and from tubs, sinks, and showers.

Humans are not the only victims. Lightning is deadly to animals too, including fish in the water. All animals inhabiting regions where thunderstorms occur face the danger. Lightning struck and killed a giraffe at Disney World in Orlando, Florida, in 2003. There have been several reports of mass deaths of herd animals attributed to lightning. In 1939, lightning killed 835 sheep bedded down for the night on the top of Pine Canyon in the Raft River Mountains of northwest Utah. Thunderstorm rain had wet the ground and sheep before the bolt hit the ground. The following morning found only 15 dazed sheep from the flock alive, along with one injured shepherd, who had been knocked unconscious. Utah was also the location of an earlier mass lightning death when, in 1918, 654 sheep were killed simultaneously in American Fork Canyon.

To stand up against the deep dread-bolted thunder

In the most terrible and nimble stroke

Of quick, cross lightning?

William Shakespeare, *King Lear*

LIGHTNING BOLTS

What they are:

In simplest terms, lightning is a large spark of static electricity in the atmosphere. It is usually connected with thunderstorms, though volcanic eruptions and wildfires can cause lightning. Lightning is always accompanied by thunder, though at times it cannot be heard.

What happens:

Lightning begins when areas of a thundercloud become electrically charged, building large voltage differentials between one section of a cloud and another, between one cloud and another, or between a cloud and the ground below. To relieve this differential, a weak, negatively charged streamer of ionized gas moves out in search of the path of least resistance to an area of positive charge. When this negatively charged stepped leader finally connects with an area of positive charge, a lightning bolt races back up the leader channel to relieve the electrical pressure. Several flashes are usually required in a single bolt to relieve all the potential charge stored in the cloud region. The complete lightning flash typically lasts about $\frac{1}{10}$ of a second.

Where to see them:

Lightning occurs anywhere thunderstorms rage. Storms are most common in the tropics and their likelihood diminishes with latitude, becoming rare at polar latitudes.

When to see them:

Although they are most frequent during the warmer periods of the year, storms can occur during cold weather and can even be associated with snowfall. The daily peak time for thunderstorms is late afternoon, when solar heating has peaked, but they can occur at all hours of the day.

A schematic diagram of electrical charge distribution in a thundercloud and the likely path of the resulting lightning bolts.

region of positive charge concentration

region of negative charge concentration

discharge between positive region of one cloud and negative region of another

discharge within cloud from negative base to positive top

typical cloud-to-ground negative discharge

Yeller gal, Yeller gal, flashing through the night,
Summer storms will pass you, unless the lightning's white.

American folk saying

POSITIVE LIGHTNING

Most of the lightning bolts that strike the ground move electrons (negative charge) down their channel. We call this "negative lightning," or just "lightning." A small percentage (less than 10 percent) of lightning, however, carries a positive charge to the ground. Termed "positive lightning," it contains much more energy—six to ten times more—than negative lighting and often strikes across great distances—as much as 10 to 15 miles (16–24km) from the cloud.

Positive flashes appear to originate in the positively charged cumulonimbus anvil region. From there, the leader moves out horizontally for several miles before becoming vertical, often connecting first with another cloud before striking the surface. Because of the great distance from the originating cloud, positive lightning can appear to strike from a clear sky, becoming the legendary "bolt from the blue"—a lightning striking during fair weather. To jump across such great distances, the voltage difference must be substantially more than for a negative bolt.

Positive lightning appears to occur more frequently during heavy thunderstorms, particularly after much of the initial fury has ended, and therefore caution must be exercised when resuming outdoor activities after a storm's passage. It also appears to occur more frequently in winter thundersnows. Because positive lightning has significantly more energy (current) and lasts about ten times longer than regular lightning, it has been blamed for starting many forest fires and causing extensive damage to electric power grids.

HEAT LIGHTNING

As the old American folk saying (above) warns, there is a difference between lightning that appears white and that which appears yellow. "Yeller gal" refers to a special case of lightning known as "heat lightning." Folk mythology incorrectly suggests it is caused by hot air expanding until it sparks on sultry summer nights.

In truth, heat lightning is not a unique lightning form, but normal thunderstorm lightning that flashes too far away from the observer for its thunder to be heard. We see heat lightning mostly at night, when darkness accentuates the visibility of "yeller-gal" lightning.

During sultry weather, scattered, short-lived thunderstorms may pop up across a region, driven by heat and humidity. Some pass a long distance from observers, and, being scattered across the sky, they require a long line of sight through the mostly clear air to be seen. When viewed from such distances, the lightning usually takes the form of sheet lightning, a standard intra-cloud or cloud-to-cloud lightning bolt seen reflected off cumulonimbus towers or diffused by the air and losing its distinctive bolt pattern. Because of the distance traveled between the bolt and the viewer, much of the bolt's blue light waves are scattered out leaving only the yellow—hence "Yeller gal." Closer lightning bolts remain white-colored, and their thunderstorms are more likely to pass close enough to drop rain nearby.

Within distant thunderstorms, the lightning flashes send their light as many as 100 miles (160km) out, depending on the height of the bolt within the cloud and the clarity of the intervening air. Yellow-tinted lightning can even be seen at great distances when flashing from towering cumulonimbus whose bases lie below the horizon. Thunder, by comparison, has a much shorter range of detection. The combined effects of reflection, refraction, scattering, and attenuation limit the distances from which thunder may be heard by a ground-based observer, to usually no more than 15 miles (24km) in quiet rural settings and five miles (8km) in noisy urban environments.

DRY LIGHTNING

A thunderstorm usually means a violent shower of rain, perhaps with hail, thunder and lightning, and gusty winds. Not all thunderstorms, however, are wet. Of course, they are not thunderstorms without thunder and lightning. Even without precipitation, however, the winds and lightning still remain. Such conditions worry those who fight wildfires because the lightning associated with such storms can spark a blaze, and since no rain falls, these lightning-initiated fires do not self-extinguish.

Sheet lightning in the background illuminates cumulonimbus towers, while cloud-to-cloud "streak" or "forked" lightning is visible on the right.

Because no rain reaches the surface, the associated lightning is termed "dry lightning." That is not to say, however, that rain is not falling from the cloud. What has happened to the moisture? Quite simply, it evaporated on its descent to the surface. We can see evidence of this below the cloud base as streaks or wispy plumes, known as *virga*. How far below the cloud the virga extends depends on the dryness of the air below the cloud base and the size of the raindrops. Small drops falling through dry air evaporate quickly. Even hail can melt and evaporate before reaching the surface. In arid areas such as western North America, from the U.S.–Mexico border north into the Canadian Prairies, rain falling from thunderstorms often evaporates completely before reaching the ground.

Ball Lightning

No other weather phenomenon has engendered more speculation or controversy than ball lightning. Only in the past half-century has its existence been slowly accepted by scientists. Sightings of ball lightning have been chronicled for centuries, including by such notables as the Roman philosopher Seneca, Pliny the Elder, Emperor Charlemagne, King Henry II of England, Czar Nicholas II of Russia, and Nobel-Prize winning physicists Niels Bohr and Pjotr Kapitza. Although ball lightning has been part of folklore for millennia, and despite more than 10,000 written accounts over the centuries, most reports have been dismissed as fantasies, illusions, or hoaxes.

Reluctance to accept its existence has a lot to do with ball lightning's rarity, variability, and unpredictability. It has been generally described as an independent spherical object, which varies in size, from that of a pea to several yards (meters) across. Reports of ball lightning have attributed a wide range of properties and behaviors to the phenomenon. While most manifestations have been described as spherically shaped, others have been seen as oval, teardrop, rod-, or disk-like in shape. Appearances range from transparent to translucent to multicolored, usually glowing as bright as a 100-watt bulb. Their motion has been described as up and down, side to side, hovering, and either moving with the wind or against it, often in erratic trajectories. When they dissipate, these balls have suddenly vanished, faded away, or exploded loudly and forcefully. Impacts on materials and living creatures have also varied widely, from safe and non-consuming to deadly and destructive. Most sightings report that they last many seconds, unlike microsecond flashes of lightning, and travel, sometimes bouncing, across the ground or other surfaces. Ball lightning has not always been associated with thunderstorms, having appeared even in calm weather.

The cause of ball lightning remains a great mystery. As many as one hundred theories have been proposed to explain ball lightning. Several theories consider it a plasma, or ionized cloud, of electrically charged particles that glow. Another group suggests it arises from combustion of materials heated by lightning bolts. A few believe it to be a glowing electrical field generated by microwave radiation. Even minute black holes have been put forward as the cause.

Soil silicon plays an important role in one currently favored theory of ball lightning formation. It suggests the orb arises when lightning strikes soil silicon, forming a vapor. This vapor then condenses into particles that slowly burn in the air's oxygen, a process that releases chemical energy.

Laboratory experiments by physicists Antonio Pavão and Gerson Paiva of the Federal University of Pernambuco, located in Recife, Brazil, have created electrical orbs from silicon vapors that mimicked many characteristics of natural ball lightning. Their diffuse golfball-sized spheres spun, sparked, and vibrated while moving erratically across the lab. The orbs rolled around the floor, bouncing off objects, and burning whatever they touched. The spheres had a lifespan of two to five seconds, with the longest lasting eight seconds. As yet, no other material has formed laboratory ball lightning, although the scientists believe aluminum and iron could also seed natural ball lighting, based on observations.

 An illustration of an indoor ball-lightning incident—an event which has been described as a "globe of fire."

GLOBE OF FIRE DESCENDING INTO A ROOM.

Fire and Brimstone

A raging thunderstorm with incessant lightning can inspire fear and awe in anyone caught in its fury. Combine a lightning storm with a wildfire or a volcanic eruption, and images of the apocalypse immediately arise. The dangerous conditions surrounding lightning formed amid fire and brimstone make conditions for field research even more hazardous for scientists seeking to unravel the mysteries of this species of lightning.

Nearly all lightning that strikes around the world is formed by thunderstorms that are raging within the great cumulonimbus clouds every day. Most thunderstorms develop from rising air masses that result from unequal air densities, high moisture content, and fast-moving winds. Two forms of lightning have completely different origins, which we can term "fire and brimstone": thunderstorms generated by the fierce heat of wildfires and by ash clouds exploding out of erupting volcanoes.

Pyrocumulus Lightning

The heat generated by wildfires, when combined with moisture pulled from forests, grass, and the soil, and evaporated from any standing water bodies, can initiate extremely fast-rising columns of air above the fire zone. At first, these columns mingle with the smoke plume above the fire, perhaps forming small smoke clouds, but if the environment is favorable, these columns explode into towering pyrocumulonimbus clouds (*pyro* meaning "fire"), sporting lightning, gusty winds, and some rain. The resulting pyro-thunderstorms expand quickly, drastically changing the conditions facing firefighters and sometimes igniting lightning-sparked fires away from the main fire. Firefighters term this very dangerous situation a "blow-up." Any rain falling from a pyro-thunderstorm usually evaporates quickly in the hot dry air within the fire.

Pyro-thunderstorms are feared because the rapidly changing conditions associated with them alter the nature of the fire and often dramatically change the direction in which the fire moves. Inflow winds associated with the thunderstorms add oxygen to the mix, while the gusty winds in the storm's circulation can stir the fire and suddenly push the flames in another direction, often causing them to jump ahead as burning embers that ignite new fires. Such conditions can trap firefighters and are a prime cause of death and injury among them.

The worst fire phenomenon is the firestorm. Firestorms combine the elements of the hottest fires with dangerous aspects of severe thunderstorms, creating something akin to hell on Earth. Some firestorms spawn fire tornadoes, whirlwinds of fiery gas blowing in excess of 150mph (240km/h),

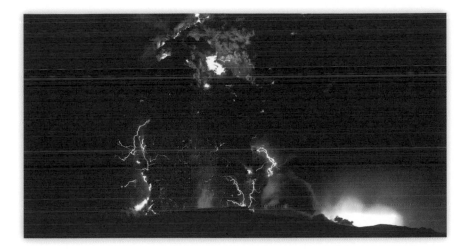

that are capable of lifting large logs. Trees ahead of firestorms can even explode into flames as if torched by unseen hands.

VOLCANO LIGHTNING

"Brimstone"-generated lightning can be seen flashing from ash clouds as they spew from an erupting volcano. The 1980 eruption of Mount St. Helens in Washington State produced a lightning bolt every second—though not all volcanic eruptions generate lightning. The processes that form volcanic lightning are less well understood than those forming thunderstorm lightning, because of the difficulty in approaching the eruption clouds.

Scientists now recognize two forms of volcanic lightning. The first occurs in the volcanic smoke and ash plume shortly after the eruption ends. The process begins, probably in a similar way to thunderstorm lightning, with the electrification of particles—but in this instance, of ash particles within the expanding eruption cloud. Aerodynamic processes segregate positively charged particles from negatively charged ones in the air above the volcano. Eventually, the charge that has built up between these regions explodes, sending out a lightning bolt, often into the clear, surrounding air.

A newly defined second volcanic lightning type arises during the volcano's explosive phase as electrically charged magma, ash, and rock leave the volcanic cone. The result produces continuous, chaotic sparking at the cone summit. How these ejecta become charged remains a mystery.

Whether eruptions sport lightning or not depends on several factors. Some volcanoes initially explode, tearing off snow-covered summits or bursting from undersea vents, and thereby producing charged particles as water and ice boil away in the eruption's heat. Studies have shown that the meeting of seawater and molten lava produces charged particles in the air above.

Lightning flashes from the ejecta cloud during the 2010 eruption of the Icelandic volcano Eyjafjallajökull.

Thunder: Voice of the Heavens

A low rumble breaks out across the landscape, far off a first, but rolling ever closer. A massive explosion instantly follows a great flash, shaking the walls and hurting the eardrums, its echo continuing for many seconds as thundercloud responses dot the sky. Thunder . . . the heavens have spoken.

Thunder, one of the most recognizable natural sounds, lends its name to common weather phenomena across Earth: the thunderstorm and its associated elements—thunderheads, thunderclouds, thunderbolts, thundershowers, and thunderclaps.

Early humans believed thunder to be the voice of the deities. The roster of thunder gods includes the Vikings' Thor, the Germanic Donar, the Greeks' Zeus, the Romans' Jupiter, the Celts' Taranis, the Slavs' Perkunis, the Indians' Indra, and the Nigerian Yoruba's Shango. Each threw thunderbolts earthward while their voice reverberated across the heavens. Other cultures attribute thunder and lightning to a thunderbird, which frequently provides the basis for Native American and African tribal legends. For example, the Bantu of southern Africa believed the beating of Umpundulo's wings produced thunder as the thunderbird dove toward earth.

Many early European cultures believed thunder an omen beyond the dread of thunderstorms. For example, the Greeks believed thunder on the right was a good omen, while the Romans regarded thunder on the left as favorable. Both agreed that thunder in the east was more favorable than that in the west—perhaps because, since weather generally moves from west to east, eastern thunder indicated the tempest had passed.

Eventually, people realized that thunder had natural causes, explicable through observation and logical deduction. Today, atmospheric scientists can explain thunder and its various voices as beginning with air and lightning. Having a temperature as high as 54,000°F/30,000°C (hotter than the solar surface), each lightning flash superheats the air surrounding its path. As the superheated gas rapidly expands, a shockwave radiates from the lightning channel. This shockwave rapidly loses energy to the surrounding air and "relaxes," which produces an acoustic (sound) wave that moves out perpendicular to the lightning producing it. Although less than one percent of the total shockwave energy transforms into the acoustic wave, the

I scratch my head with the lightning and purr myself to sleep with the thunder.

Mark Twain, *Life on the Mississippi*

total energy available for that sound wave remains extremely large. Thunder explodes, therefore, as one of nature's loudest sounds. A nearby thunderclap can reach around 120 decibels, which is equivalent to being within 200 feet (60m) of a jet aircraft during takeoff. A chainsaw roars at around 100 decibels.

The strength of lightning surge currents determines the duration and loudness of the subsequent thunder. The shockwave radius at relaxation establishes the characteristic pitch (frequency) of its thunder. The more powerful the stroke, the wider the channel and the lower the resulting pitch. Thunder's pitch generally falls between 15 and 40 hertz or between 75 and 120 hertz. (For reference, an 88-key piano's lowest note vibrates at 30 hertz.)

Thunder, however, produces more than a single loud, explosive sound. It rolls and rumbles through the stormy sky; it cracks and claps. Since each lightning flash is composed of a number of large jagged segments that make up the stepped leader, and these are oriented in various ways relative to the listener, thunder generally resounds with both claps and rumbles. Thunder travels through the lower atmosphere as acoustic waves moving at around 770mph (1,240km/h). Thunder's character—its pitch, loudness, and form (a crack or rumble, for example)—depends upon the lightning flash and

A spectacular lightning display illuminates the eruption of Galungung volcano in Western Java, Indonesia in 1982.

the order in which the sound waves reach the observer. For example, if the lightning channel lies broadside to the listener rather than end-on, then most of the sound waves it creates will head toward the listener, arriving nearly simultaneously and producing a short, loud thunderclap.

The atmosphere also modifies thunder's volume, pitch, and character. Since atmospheric density varies both vertically and horizontally, and winds blow through it at various speeds and directions, thunder may be scattered, attenuated, refracted, or reflected on its way to the observer. The atmosphere's temperature usually decreases at higher altitudes, and sound travels faster in warm air than in cold, so thunder waves passing through the lower atmosphere curve upward (refract). Sound also moves faster downwind than it does upwind. Atmospheric scattering and attenuation alter the total sound reaching listeners by reducing higher frequencies, so after traveling several miles, thunder becomes a low-pitched rumble.

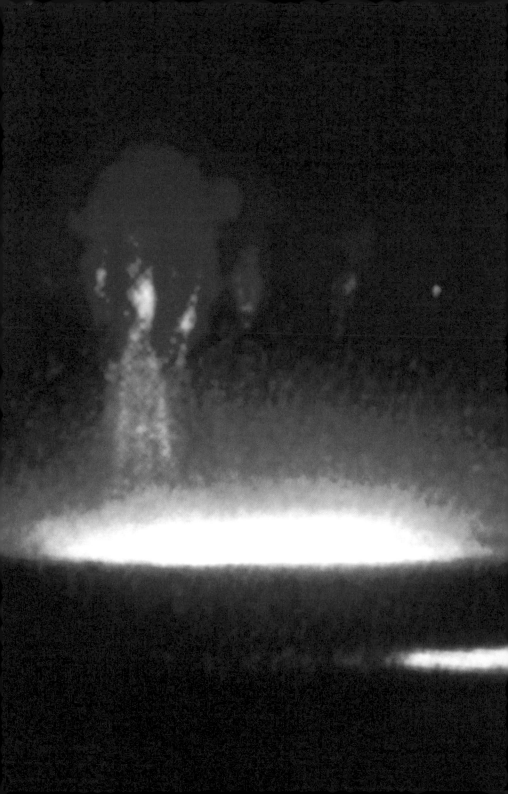

Jets, Sprites, ELVES, and Trolls

Blue jets, red sprites, ELVES, sprite haloes, and trolls may sound like the offspring of Tolkien characters born in the psychedelic sixties, but they constitute the newest members in the pantheon of atmospheric electrical phenomena, alongside St. Elmo's Fire and lightning. Although they were first reported as unidentified oddities as far back as 1886, it took nearly a century for the meteorological community to accept their existence.

The existence of jets, sprites, ELVES, sprite haloes, and trolls remained questionable for nearly a century, despite anecdotal evidence, due to their very transient nature and their occurrence above towering cumulonimbus clouds. Collectively termed "transient luminous events" (TLEs), it took the advent of manned orbital observatories and high-altitude aircraft operations to bring them into meteorological discussions.

FIRST SCIENTIFIC OBSERVATIONS

The groundwork began in the latter decades of the 19th century, when observations of mysterious flashes above thunderstorms were reported. Because they appeared to shoot upward from the cloud tops, they were often called "rocket lightning." In the earliest known report, in 1886, T. MacKenzie and M. Toynbee noted that "continuous darts of light . . . ascended to a considerable altitude, resembling rockets more than lightning." In 1903, W. H. Everett presented a paper advancing his observation of rocket lightning as having "a luminous tail . . . shooting straight up . . . rather faster than a rocket." Around 1920, Nobel laureate, British physicist C. T. R. Wilson saw "diffuse fan-shaped flashes of greenish color extending upward into the clear sky" and felt this might represent a new aspect of atmospheric electricity.

Initially called "cloud-to-space" lightning and rocket lightning, TLEs have now assumed a variety of colorful names. According to TLE-pioneer Dave Sentman of the University of Alaska-Fairbanks, the first member,

The first true color image of red sprite lightning above a thunderstorm, photographed on July 3, 1994, from an altitude of 8 miles (12.9km) over the USA. Red sprites last for a few milliseconds and are barely visible to the naked eye. They are associated with the most energetic positive-charge cloud-to-ground lightning strikes.

the sprite, was named after the mischievous air spirit Puck in William Shakespeare's play *A Midsummer Night's Dream* because of its transient, ephemeral nature.

Pilots flying over the top of thunderstorms reported many early observations of TLEs when they saw flames or lightning bolts issuing from the thundercloud tops headed toward space. Scientists from the University of Minnesota took the first recognized photo on July 6, 1989. A year later, they shot the first videotape of the phenomenon, confirming that these apparitions constituted a completely new form of lightning. After their discovery, close reviews of thunderstorm videos taken from the Space Shuttle revealed previously unknown images of TLEs, confirming the aircraft observations. Like many newly documented discoveries, the seemingly rare suddenly became commonplace as more observations from the Space Shuttle and International Space Station, high-flying aircraft, and even from ground level, soon followed. Knowledge of the properties and science of TLEs is still unfolding but there appear to be several distinct categories of these phenomena. Four of these are explained below.

SPRITES

The earliest observed sprites all had one color, and the initial moniker given was "red sprites." Further detailed observations found they also contained faint tendrils of blue and purple elements. As the observation database grew, sprites began to exhibit a wide range of sizes and shapes, including giant red blobs, picket fences, upward-branching carrots, or tentacled octopuses! Sprites can extend as far as 60 miles (100km) above the surface with downward-draping tendrils often dropping below 20 miles (30km) altitude, although they do not actually touch the thundercloud tops. The peak brightness of a sprite usually occurs at an altitude of between 30 and 50 miles (50–80km).

Instead of forming narrow channels like other types of lightning, sprites spread to about 30 feet (10m) across and often appear as large clusters that illuminate a large volume, perhaps thousands of cubic miles, spreading out over 100 miles (160km) from above very large thunderstorm systems. Sprites appear to associate with positive cloud-to-ground lightning, flashing at intervals of several seconds while lasting for only milliseconds. At times, a diffuse, disk-like glow lasting a split second precedes some sprites. Called "sprite haloes," these are less than 60 miles (100km) wide and propagate from about 53 miles (85km) down to 44 miles (70km).

The lightning strikes that precede sprites emit unique radio signals. Using this property as a detection device, researchers now believe sprites form during roughly one in every 200 lightning bolts.

BLUE JETS

Astronauts first captured this "rocket lightning" on video on October 21, 1989, as their Space Shuttle passed over Australia, although this discovery was not realized until the following year. A few years later, while flying research aircraft above severe thunderstorms in Arkansas in 1994, University of Alaska scientists Eugene Wescott and Davis Sentman witnessed blue light beams shooting upward directly out of cloud tops, confirming the phenomenon's existence. (At the time, the Shuttle's cameras weren't powerful enough to take detailed images of these phenomena.) These TLEs rose at over 60 miles (100km) per second to altitudes of

25 to 30 miles (40–50km), and then faded from view after about 100 milliseconds. Wescott and Sentman branded these flashes "blue jets" and confirmed MacKenzie and Toynbee's 1886 observations. During this flight, the color imagery established that jets were blue, and sprites were red.

Blue jets generate from thunder-cloud tops heading toward the ionosphere located 12 to 30 miles (20–50km) above the ground. Lasting from tenths of a second to a full second, the jets are always blue and assume a funnel shape 1 to 2 miles (1.6–3.2km) wide at their base and 5 to 6 miles (8–10km) across at the top. Blue jets appear to be rare, but this may be because their faint blue light quickly becomes scattered by the surrounding air, making the jets difficult to see from the ground.

At times, several blue jets simultaneously propagate slowly upward from the cloud tops, and, interestingly, they extinguish simultaneously. A related phenomenon, the "blue starter," may actually be a blue jet that fails to form completely.

ELVES

After sprites and jets had been identified in the Space Shuttle videos, a distinctly different flash event was noticed in analysis of video footage taken directly above an active thunderstorm complex off the coast of French Guyana on October 7, 1990. In 1995, scientists from the University of Tohoku (Japan) and Stanford University (USA), using ground-based observations, confirmed these were new TLE members, giant expanding disks of light appearing between 40 and 60 miles (65–100km) altitude. In keeping with the fairytale motif, they are called ELVES (an acronym for "Emission of Light and Very Low-Frequency Perturbations From Electro-Magnetic Pulse Sources"). They arise when powerful lightning flashes emit electromagnetic pulses in the form of intense radio waves. As the pulses cross the ionosphere, they excite electrons in the atmospheric nitrogen gas, causing the nitrogen to glow by fluorescence. Although huge in size, more than 250 miles (400km) in diameter, they are extremely transient, lasting less than a millisecond, so it is unlikely that the human eye can see them. ELVES need not appear close to their triggering lightning, often arising as many as 30 miles (50km) away from the initiating bolt.

TROLLS

Next, we have trolls. These TLEs have also been called the more aptly descriptive "embers" and "fingers," but to keep with the theme, "troll" is an acronym for Transient Red Optical Luminous Lineament. Trolls are similar to blue jets but are generally reddish in color. They occur following especially vig-

---

In its most typical form it consists of flames appearing to shoot up from the top of the cloud or, if the cloud is out of sight, the flames seem to rise from the horizon.

Nobel laureate C. T. R. Wilson, c. 1920

orous sprites whose tendrils have extended downward to near the cloud tops. It is still not certain whether the sprite tendrils actually extend to the physical cloud tops, or if trolls emerge from the storm cloud. Trolls have luminous heads that leave faint trails and initially shoot upward at around 95 miles (150km) per second, before gradually decelerating to vanish around 20 miles (30km) up.

Gnomes and Pixies

Finally, in keeping with the theme, two additional TLEs, gnomes and pixies, have also been identified. Gnomes are small, brief white spikes that shoot upward from the anvil-top of a large thunderstorm. Lasting

but a few microseconds, they extend upward more than 2,500 feet (800m) and spread to a width of around 450 feet (140m). Pixies are "point" objects less than 300 feet (90m) across. They were initially discovered as a number of lights flashing randomly around the thundercloud's overshoot dome.

Although thousands of thunderstorms rage across Earth every second, TLEs were not positively identified until high-altitude aircraft were able to photograph their short lives.

Lightning Rocks!

Weather beats continually on the rocks of Earth's lithosphere, eroding landscapes into wondrous shapes through processes that take millennia, if not millions of years. But one rock form arises almost instantaneously. It occurs when lightning strikes a surface high in sand content or other loose quartz materials. Popularly called "petrified lightning," the resulting object is technically known as a "fulgurite."

When a lightning bolt strikes an object, its high energy—some 300 kilowatt-hours—can burst wood into flame, melt metals and rocks, or fuse the minerals that make up some soils. As a current, lightning can penetrate the soil, heating it to extreme temperatures through the resistance of the soil components. When that interaction involves a silicon-based mineral, the product can be spectacular.

Silicon ranks as the second most common element on Earth behind oxygen, with which it combines to form silicon dioxide (SiO_2) (quartz) and silicate (SiO_4) ion—a building block for other minerals such as feldspar and mica. Pure common sand and basic glass are composed of silicon dioxide. The melting point of SiO_2 is 2,950°F (1,620°C)—much less than the temperature of a lightning strike, which can approach the temperature of the solar surface. We call silicon materials hit by lightning, melted, and then fused, "fulgurites"—from the Latin *fulgur*, which means "thunderbolt." There are two types of fulgurites that have been recorded: the sand fulgurite and the rock fulgurite.

FULGURITE FORMATION

When lightning strikes a rock surface, melting and fusing the minerals on its surface, rock fulgurites form. This fulgurite appears as a glassy crust on the rock, often with short tubes or perforations lining it. Rock fulgurites can include materials other than silicon-based minerals. The glass coating varies in color, depending on the minerals involved. Most rock fulgurites have been found on vegetation-free mountain summits that naturally attract surface lightning strikes, including Europe's French Alps and Pyrenees ranges, and the western mountains of North America, including the Rocky Mountains, the Sierra Nevada, and the Cascade and Wasatch ranges.

Sand fulgurites form in any sand or sandy soil surfaces, such as beaches and deserts. When the lightning hits the surface and enters the ground, it vaporizes and melts the silicate soil material, which almost immediately fuses into a hollow tube of fragile glass that may extend several yards (meters) deep. The nature of sand fulgurites depends on the composition of the struck soil. The best specimens form in clean quartz sand. Their color depends on the sand composition and

ranges from translucent white, through tan and green, to black.

The most favorable condition for sand fulgurite formation is a relatively dry quartz sand layer with a more electrical-conducting layer or the water table just below. Their formation takes around one second. The lightning strike rapidly heats the air and moisture present in sandy soil. The explosive expansion of the soil gases pushes molten quartz outward where it rapidly solidifies around the central void.

Fulgurite tubes, sometimes referred to as petrified or fossil lightning, generally have diameters of one to three inches (2.5–7.5cm), decreasing with depth. The

inner surface of the tube is smooth and glassy, perhaps with surface bubbles or blisters of trapped air. Tube exteriors are rough, often covered with adhering sand grains. Their length can extend to several feet (1 meter or more), at times taking a root-like appearance with several branches off the main tube. The record fulgurite recognized by the *Guinness Book of Records* measured 17 feet (5.2m) long and was excavated in northern Florida in 1996. It was composed of two vertical branches, the second being 16 feet (4.9m) in length.

Because fulgurites are made of glass, they are extremely resistant to weathering, and therefore may be preserved for millions of years. One fossil fulgurite has been estimated to be 250 million years old. Consequently, they have been used as indicators of paleoenvironments through the study of the composition of the gases within the wall bubbles. Fulgurites have been used as a folk cure for a number of ailments and in crystal healing, and as a charm to ward off lightning.

 These fulgurite segments, found in the Sahara desert in Mauritania, are the result of lightning striking a sandy surface.

Will o' the Wisp: The Fairy Lights

Are they electrical? Are they chemical? Are they real?
One group of mysterious atmospheric phenomena has been
observed by peoples across the world for thousands of years,
and they go by a variety of names including "will-o'-the-wisp,"
"the min-min lights," "ghost lights," "spunkies," and the "hobby
lantern." Adding to their mysterious nature, these lights
appear to favor swamps, bogs, and marshes.

Walking through a marsh, bog, or swamp at twilight or night, you see an eerie light flickering like a lamp behind you . . . or ahead . . . or to the side. What is it? Who's out there? Is this something supernatural? Such mysterious lights have been part of folklore for centuries. In English-speaking countries, they go by many names including will-o'-wisp, jack-o'-lantern, friar's lantern, Hinky Punk, and the min-min lights. The most common name, "will-o'-the-wisp," comes from the name *will and wisp* meaning a torch or light. The Latin name is *ignis fatuus*, meaning "foolish fire" (*ignis* meaning "fire" and *fatuus* meaning "foolish"). In most cases, these lights do not carry a good reputation.

Across Britain, there are regional names for it: spunkie in the Scottish lowlands; the hobby lantern in East Anglia; Peg-a-lantern in Lancashire; Hinky Punk in Devon; and Ellylldan in Wales, to name but a few. Australians have the min-min lights; the United States, the ghost lantern; the Germans, *Irrlicht* ("foolish or errant light"); and the Dutch, *dwaallicht* ("wandering light"). In all cases, the lore attached to these lights

centers around spirits drawing the viewer to follow their lights (lanterns) into the unknown. Most stories indicate appearances around peat bogs, marshes, and swamps; however, the ghost light or corpse fire appears in graveyards as an omen of impending death. The *Irrlicht*, described as a ball of fire, lures people off the road and into the woods, while the min-min lights, aka "dead men's campfire," are believed to be the spirits of lost or stillborn children who seduce the viewer into the desert, never to return.

Will-o'-the-wisp lights have been described as faint, flickering, moving erratically, and never approachable. When considered a "flame," it is non-consuming and dim. Its colors have ranged from blues and yellows to reds and greens.

Several theories have been put forward as scientific explanations. One suggests that they are a form of ball lightning produced by a piezoelectric effect, an electric field generated by crystals or ceramic materials under pressure. The pressures in these events are believed to result from tectonic strains in the earth's crust on near surface quartz, silicon, or arsenic rock. A form of St. Elmo's Fire could also explain some sightings.

Another theory suggests they result from bioluminescence, which is naturally occurring and generated by biochemical materials that glow with a cool, non-consuming light. A number of fungi and insects exhibit bioluminescence, including the firefly, which uses its bioluminescent tail as a sexual attractant.

A third theory, proposed by Professor Jack Pettigrew of Queensland University, suggests that they are associated with the superior mirage, similar to the Fata Morgana, causing distant lights to take on a magical quality. One point in favor of this explanation is that the lights can never be approached—they advance and recede counter to the motion of the observer, which is a property of mirages.

The most accepted explanation cites the low-temperature combustion of methane gas, a decay gas common to marshes and swamps, assisted by phosphine released by the bacterial reduction of phosphate

in decaying organic matter. Controlled experiments indicate faint luminescence only requires small quantities of either gas. The gases can be emitted erratically over a marsh-like terrain, explaining the transient and spontaneous nature of these lights. The low combustion temperature also explains the non-consuming nature of the fire.

The will-o'-the-wisp family remains mysterious. They may form through one of the above explanations, or perhaps by none of them. Alternatively, each explanation may be correct depending on the local environmental conditions when the ghostly lights appear in the gloaming.

A detail from an engraving by Josiah Wood Whymper from 'Phenomena of Nature'(1894) showing a Will-o'-the-Wisp glowing among the reeds in marshy land alongside a stream.

Ignis fatuus, called by the vulgar Kit of the Candlestick, is not very rare on our downes about Michaelmass . . . a point of light, by the hedge, expanded itselfe into a globe of about three inches diameter, or neer four, as boies blow bubbles with soape. It continued but while one could say one, two, three, or four at the most. It was about a foot from my horse's eie; and it made him turn his head quick aside from it. It was a pale light as that of a glowe-worme.

John Aubrey, *The Natural History of Wiltshire*, c.1697

[Will o' the Wisp, or Ignis Fatuus] appeared in the same
place and was visible quite half-an-hour. It does not dance
about, but now and then takes a graceful sweep, now to quite
a height, and then makes a gentle curve downwards, after
sparkling and scintillating away for ten minutes or more.

F. Ramsbotham, "The 'Ignis Fatuus,' or, Will o' the Wisp," in the

Quarterly Journal of the Royal Meteorological Society, 1891

CHAPTER FIVE

geological

phenomena

So far we have looked at natural wonders whose origins
are in the heavens, in the skies, and in the atmosphere
around us. In this chapter we investigate the phenomena
that emanate from the ground beneath our feet. Some are
artistic, some are life-threatening, and all are awe-inspiring.

The Shaking Earth

The ground we walk upon seems solid enough, and most of the time it is, but Earth's crust is just that—a thin, cool crust floating on the surface of a ball of red-hot molten rock. Everything on this planet is on the move, pushed and pulled by unimaginably huge forces. Their actions are usually imperceptible, but when these forces burst through to the surface, the results can be catastrophic.

As one of the most powerful events that ever takes place on our planet, an earthquake is undoubtedly a natural wonder, but one we hope we will never experience. A major earthquake can take literally thousands of lives, so what causes these earth-shattering phenomena, and why are certain parts of the world earthquake-prone?

Earth's hard, rocky outer layer, or crust, consists of several separate pieces known as tectonic plates, and these correspond broadly to the continents and the associated ocean bed. The average thickness of the plates is about 20 miles (35km) but they can be up to 60 miles (100km) thick beneath the major mountain ranges. The crust that forms the ocean bed is much thinner, averaging only 3 miles (5km) in thickness.

These tectonic plates, which have solidified by cooling, are literally floating and they are constantly moving relative to each other, driven by thermal currents and upwellings in the molten layers below, which remain in a liquid or semi-liquid state because of heat retained from the time of the Earth's formation and energy generated by the radioactive decay of elements within the Earth.

Earthquakes occur at what is called a "transform boundary," the frontier between two tectonic plates where one plate is moving past another. Rather than sliding easily past each other, the rough edges of the plates can become locked together, and the continued movement of the two huge plates causes the crust at the boundary to deform, leading to a huge buildup of stored energy. When the two plate edges finally and suddenly move past each other, this stored energy is released in the form of seismic shock waves that travel out from the point

The Abruzzo region of central Italy lies in a tectonically active zone. When an earthquake struck in 2009, many of the medieval villages that surround the regional capital of L'Aquila were literally flattened, causing more than 300 deaths, and prompting an international rescue operation. In 2010, a devastating earthquake hit Haiti, killing thousands and creating an ongoing humanitarian crisis.

An earthquake achieves what the law promises but does not in practice maintain—the equality of all men

Ignazio Silone, Italian author and politician

of movement, known as the "focus," causing the crust to shudder and undulate, often over a large area. The energy released by a quake can bring buildings and bridges—even entire cities—crashing to the ground, send giant waves smashing into coasts (see p. 202), break dams, divert rivers, and cause landslides that transform the landscape. The loss of life can be huge. The earthquake that hit the Caribbean nation of Haiti in January 2010, with its focus close to the capital city Port-au-Prince, is believed to have killed some quarter of a million people and made a far greater number homeless. Haiti lies on a transform boundary between the North American and Caribbean plates.

The west coast of California is renowned for its earthquakes, and the reason for this is that the 810-mile (1,300km) San Andreas Fault, another transform boundary, runs through the area. Here, the Pacific tectonic plate is moving northward relative to the North American plate. The cities of Los Angeles, San Diego, and Santa Barbara are actually located on the edge of the Pacific Plate, while San Francisco, which suffered enormous damage in the great earthquake of 1906, is on the North American Plate. Although the rate of "slippage" is less than one and a half inches (3.7cm) per year, Los Angeles will eventually move north of San Francisco.

VOLCANOES

The Pacific Plate is notorious for having very active boundaries with the plates that it abuts on all sides. As well as the transform boundaries that characterize parts of the American west coast, on its northern and western edges it has what are called "convergent boundar-

 The earthquake that shook San Francisco in the early hours of April 18th, 1906, was felt as far away as Oregon and Nevada. More than 3,000 people died, many of them as a result of the fires that raged through the city.

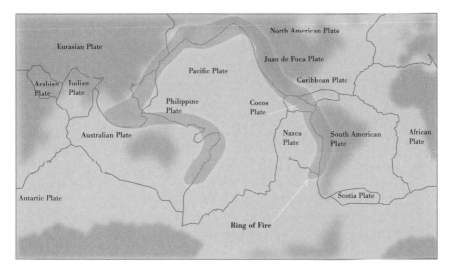

The majority of the Earth's volcanic activity takes place on the so-called Ring of Fire, along the edges of the Pacific plate and the plates that abut it.

ies," where plates are moving toward each other, and these give rise to another high-energy natural wonder: volcanoes.

Indeed, volcanoes are so numerous around the Pacific Plate that its perimeter is known as the Ring of Fire. At a convergent boundary, one tectonic plate is being pushed beneath another as the two collide, a process known as "subduction." This causes the downward-moving edge of the plate to become molten, while the edge of the other plate tends to buckle. Fissures form in the upper plate, and the molten rock, or magma, can then be forced up through the crust under enormous pressure, together with ash and gases. The vent or rupture becomes a volcano as the solidifying lava (as magma is called once it reaches the surface) forms a steep-sided cone with a crater at the top through which the molten rock emerges.

Most of the world's most violent and destructive volcanic eruptions have taken place around the Pacific Rim, including the eruption of Mount Tambora, Indonesia, in 1815, which killed more than 10,000 people with its initial explosion and affected the weather so severely throughout northern Europe and parts of North America that it caused localized famines the following year. The nearby island of Krakatoa exploded in 1883 with almost 13,000 times the power of the nuclear bomb that destroyed Hiroshima, blasting away some 5 cubic miles (21 cu km) of rock and debris, and creating a series of huge tsunamis (see page 202). On a much smaller scale, Mount St. Helens, in Washington State, USA, killed 57 people when it erupted in 1980, demolishing a substantial portion of the summit and reducing its height by some 1,300 feet (400m).

Volcanoes can also form at divergent boundaries, where the tectonic plates are moving apart. This occurs along the mid-Atlantic ridge, for example, giving rise to the volcanic islands of Iceland.

HOT SPRINGS AND STEAM VENTS

In areas of volcanic activity, hot molten magma is relatively close to the surface, and this heat energy can give rise to some dramatic "hydrothermal" (hot-water) spectacles. Rock in this kind of geological formation, which is generally volcanic in origin, is often fairly porous and contains fissures and cracks as the result of movements in Earth's crust.

Water in the form of rainfall or snow melt can make its way down through the crust until it approaches the level of the molten magma, possibly thousands of feet below the surface. Here it is heated up by the energy in the hot rocks, and the water can reach boiling point or even higher temperatures—up to 400°F (200°C)—becoming superheated as a result of the pressure from the weight of water above it. As the heated water returns to the surface, it cools sufficiently to emerge from the ground below its boiling point in the form of a hot spring, and it may flow up into a pool or lake, raising the water temperature well above that of the surrounding environment. Some hot springs pour out hundreds of gallons of hot water every minute.

Areas of volcanic and geothermal activity are often rich in a form of rock called "rhyolite," which contains silica—an ingredient of glass. This silica becomes dissolved in superhot water, and some of this mineral is deposited when water evaporates from the hot spring, creating distinctive rock formations.

The dramatic activity of geysers such as Old Faithful in Yellowstone National Park demonstrate the vast reserves of thermal energy that lie beneath our feet.

VOLCANOES

What they are:

Volcanoes are places on the globe where molten and semi-molten rock, together with gas and ash, erupt through the Earth's crust.

What happens:

The most explosive types of volcano form where one of the Earth's tectonic plates is colliding with another and being forced down beneath it, which causes its edge to melt. The resultant magma, which rises through weaknesses in the continental crust, tends to be thick, trapping gases and causing a buildup of pressure that results in an explosive eruption through a vent. Successive eruptions lead to the formation of a stratovolcano, a layered cone of ash and solidified lava. Less violent volcanoes form at divergent boundaries, where tectonic plates are moving apart, usually beneath the ocean, and allowing much thinner magma to flow out between the plates. Occasionally, volcanoes can form as a result of "hot spots" beneath the ocean floor. The Hawaiian island chain formed in this way.

Where to see them:

With the exception of those beneath the ocean, almost all volcanoes form at "destructive" convergent boundaries between tectonic plates, such as around the perimeter of the Pacific plate, through the Malay archipelago between south Asia and Australia, and in the Mediterranean, where the African and European plates are colliding. Iceland, which has more than 100 volcanoes, appears to be on both a divergent boundary and a hot spot.

The majority of volcanoes occur where the edge of one tectonic plate is moving beneath the edge of another, causing the first to melt and the second to buckle, creating a weakness through which the magma can erupt.

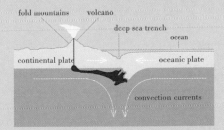

In some springs, water that is superheated to a temperature above the natural boiling point can reach the surface of the spring. When it does so, the drop in pressure causes it to turn rapidly to steam, producing a churning effect in the water. This can be seen in several of the hot springs in Yellowstone National Park in northwest Wyoming, a volcanic area centered on North America's largest volcano and containing half of the world's geothermal features. Here, too, there are mud pots, springs in which the water contains sulfuric acid. The acid dissolves the rock through which the water rises, and the spring is a bubbling pool of thick, hot mud.

Some springs contain so little water that only steam emerges from the ground, creating a steam vent, or "fumarole," that sends a steady, hissing plume of water vapor up into the air.

Geysers

The silica that dissolves out of the rhyolite plays an important role in the creation of the most spectacular of geothermal wonders—the geyser. As the silica-rich superheated water makes its way up to the surface, silica is gradually deposited on the walls of the channels and tunnels, creating a watertight

"plumbing system" through which the water travels, and this gradually constricts the flow close to the surface. This increases the pressure in the system and allows the water to remain at a temperature well above boiling point in a reservoir beneath the surface. As the temperature increases further, pockets of water at the bottom finally turn to bubbles of high-pressure steam that rise to the top of the reservoir. When these force their way out of the constricted opening there is a sudden drop in pressure below, and the superheated water beneath instantly boils, causing a jet of steam and water to explode out of the ground. Unless they erupt from within pools, geysers gradually build up a conical structure of deposited silica, known as "geyserite" or "sinterite," around the opening.

There are estimated to be some 1,000 geysers in the world, and Yellowstone Park contains about half of them. The park's most famous geyser, Old Faithful, shoots thousands of gallons of water more than 185 feet (56m) into the air approximately once every 90 minutes—hence its name. Yellowstone is also home to the world's largest geyser, called Steamboat, whose giant hot-water fountain can reach a height of 400 feet (120m), though it erupts infrequently.

Judge, then, what must have been our astonishment, as we entered the basin at mid-afternoon of our second day's travel, to see in the clear sunlight, at no great distance, an immense volume of clear, sparkling water projected into the air to the height of one hundred and twenty-five feet.

Nathaniel P. Langford, 1871

G E Y S E R S

What they are:

A geyser is a fountain of steam and water that erupts from the ground under pressure from a reservoir of superheated water below.

What happens:

Groundwater seeps down through volcanic rock to levels where it is heated by the magma below. Rising up through a natural plumbing system of cracks and tunnels, the water deposits silica that constricts the channels and allows pressure to build up in a chamber below the surface. When a certain pressure and temperature are reached, some steam escapes from the geyser's vent, the pressure in the chamber drops causing the water to turn to steam, and the geyser erupts.

Where to see them:

Geysers tend to occur in groups or "geyser fields" where the underground heat and volcanic rock formations produce the right conditions. In addition to Yellowstone Park, geyser fields are found in various locations around the globe, including: the Aleutian island of Umnak, Alaska; the Kamchatka Peninsula in Siberia; El Tatio in Chile; New Zealand's North Island (where the Waimangu Geyser was the world's largest geyser, occasionally erupting to a height of 1,640 feet [500m], until a landslide altered the underground conditions); and Iceland, home of the original Geysir, meaning "gusher" in Icelandic.

When to see them:

Geysers vary in the periods between eruptions, from every few minutes to once in several years, but if you visit a geyser field you are bound to see an eruption.

Geysers are formed where water is heated by the presence of magma beneath volcanic rock. The deposition of silica creates a plumbing system in which pressure can build up.

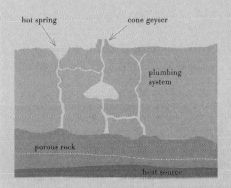

hot spring

cone geyser

plumbing system

porous rock

heat source

Erosion

Just as a sculptor's tools can cut, carve, shape, and smooth a range of materials to create artistic forms, so the forces of wind, water, ice, and gravity work upon the geological materials of the landscape and create objects of surprising architectural beauty.

We have seen how movements of Earth's crust and the molten rock on which it floats can bring about sudden, and even explosive, changes to the landscape, but erosion works its magic in much slower and more subtle ways. It also requires softer material to work with. The kinds of rock that form from magma or lava are generally extremely hard. They are known as "igneous" rocks, and they include such rock as granite and rhyolite. Much of Earth's crust is composed of this kind of rock, although the surface layers generally are not.

The majority of Earth's land surface bears layers of rock formed from sediment that has been deposited and compressed over millions of years. These "sedimentary" rocks are much softer and are more readily worn away by the action of wind and water, and where they are juxtaposed with the much harder igneous rocks, the resulting sculptures can be breathtaking.

RAW MATERIALS

Imagine a vast plain of sedimentary rock that has been laid down in successive layers over millions of years, each one containing different kinds of sediment and different minerals, creating bands of different colors and varying degrees of hardness. Volcanic activity might add layers of harder rock into the mix. Over time, movements of the Earth's crust cause sections of this to be lifted up, breaking away from the surrounding rock, possibly tilting as sections are pushed together. And now the wind and the rain begin their work.

This is the story behind the dramatic landscapes to be found in many, usually arid, areas of the world, from parts of Australia, India, South Africa, and Spain to the southwest of the United States and the Badlands of Canada. The general term "badlands" has been applied to many of these areas because they are so difficult to travel through, the terrain being broken up by high plateaus and deep gorges. Let's look at some of the features that are found here, and how they came to be this way.

Eroded by water, the layered red and orange rock of Bryce Canyon, in Utah, has been sculpted into strange pinnacles known as "hoodoos." Their irregular shape, which distinguishes them from other vertical rock formations such as spires, has helped to make them popular tourist attractions.

In fact, just about all the major natural attractions you find in the West—the Grand Canyon, the Badlands, the Goodlands, the Mediocrelands, the Rocky Mountains, and Robert Redford—were caused by erosion.

Dave Barry, American writer and humorist

MESA, BUTTE, AND SPIRE

Although the landscape we are discussing is generally found in arid areas, water is the chief cause of erosion. This is because, when rain does fall, it comes in a sudden, heavy downpour and flows from the higher land with considerable force. Because the rain is infrequent, there is little plant life to hold the water or bind the soil together. Flash floods make their way down through fissures in the rock and carry away the softer rock, and over long periods of time this carves the edges of our sedimentary plateau into separate blocks of rock. The surface of each block is also eroded by the falling rain and by wind that scours away the rock, but where a layer of harder rock becomes exposed it acts as a "caprock," preventing further erosion of the rock beneath it. Water flows off this resistant rock layer and eats away at the sides of the block below. Together with the abrasive action of wind-borne dust and sand, this creates a flat-topped, straight-sided feature that the first Spanish explorers called a *mesa* (pronounced "maysa"), meaning "table," when they came upon the stunning landscape of the American Southwest. Because the rock is made up of successive layers of sediment—commonly sandstone—laid down over long periods of time and each containing a different range of minerals, erosion reveals these layers as bands of red, orange, yellow, gray, and brown, giving

the landscape a distinctive appearance, especially in the warm light of sunrise and sunset.

Frost also plays a role: water enters crevices and cracks in the rock and then freezes in cold periods. As it does so, it expands, and in a process known as "ice wedging" it can break away large sections of rock. Together, these processes undercut the flat caprock and cause pieces of the harder surface layer to break away and fall. The mesa gradually becomes narrower, and a bank of fallen debris builds up at the bottom, giving the base a slight curve. When the mesa becomes narrower than it is tall, it is referred to as a *butte* (pronounced "byoot"), a term that is derived from the Old French word meaning "mound."

With continued erosion the butte becomes narrower and narrower, forming an ever more slender spire or pinnacle until, eventually, the erosion-resistant caprock tumbles and the wind and rain are free to erode the spire away completely.

The towering blocks of this eroded red rock fin formation in Arches National Park near Moab, Utah, have earned it the name of Wall Street.

HOODOOS

While mesas, buttes, and spires have fairly straight vertical sides, there is a far more sculptural feature of the mesa landscape that occurs when the layers of rock have varying degrees of hardness and are eroded to different extents. This happens especially when sedimentary layers are topped by compacted volcanic ash and then by a hard capstone. Water and wind-blown sand work away at the softer lower layers, creating intricate contours, while the structure remains stable under its well-supported capstone. The bizarre and otherworldly forms that result are called "hoodoos," "fairy chimneys," or "tent rocks," and they can be as tall as 100 feet (30m), though most are just a few yards (meters) in height. Very small hoodoos are often referred to as "hoodoo rocks." Some hoodoos are extremely asymmetrical, and look as though they will topple at any moment, but they can last for thousands of years if the caprock remains undisturbed. Hoodoos are found in several parts of the world, including Turkey, Taiwan, and the badlands east of Drumheller in Alberta, but the greatest number are found in Bryce Canyon National Park, in Utah, USA.

STONE ARCHES

Utah is also home to the world's greatest concentration of natural stone arches. These have been formed by the erosion of sandstone "fins," long walls of sandstone that remain when the rock on either side has been broken up by ice wedging and then washed away. Over time, the fins themselves begin to break up, but when there is sufficient strength in the upper layers the lower parts can fall away to leave spectacular arches and bridges.

Arches National Park, where there are almost 2,000 natural rock arches, contains the world's largest sandstone bridge. The graceful curve of Landscape Arch has a span of 290 feet (80m), but it continues to erode and, after several major rock falls in the 1990s, the Park Service closed the trail that passes beneath it for the sake of safety.

Bryce Canyon and Natural Bridges National Monument, both in Utah, also have

many natural arches, but these remarkable features can also be seen in the desert regions of Israel, in the Tirumala hills in Andhra Pradesh, India, and in the Czech Republic.

Not all natural arches are made of sandstone. In many coastal locations, and occasionally over water courses, arches can also form in limestone, a sedimentary rock that is largely composed of the skeletal remains of small marine creatures such as corals and foraminifera. Where limestone is exposed on a coast, near-vertical cliffs form through a combination of wave erosion at the base and weathering of the top edge that causes sections of the rock to fall away over time.

Caves are formed when waves, and the abrasive mineral material that they carry, force their way into cracks in the cliff face and grind the rock away. A headland is formed when the sea erodes sections of softer coastal rock on both sides of a band of harder rock, creating a bay on each side. When a cave forms in a headland, it may eventually break through and become an arch. A good example is seen in Australia at the Port Campbell National Park in Victoria, where London Arch consisted of a double bridge until 1990, when one span collapsed. Another famous limestone arch, known as Durdle Door, is found in Dorset, England. When coastal arches like these finally collapse from erosion, the resulting pillars are

known as "stacks," and these are worn down over time to form stumps. The Needles, off the coast of the Isle of Wight in England, are chalk stacks, while the Old Man of Hoy, a 450-foot (137m) pillar on the Scottish Orkney Islands, is composed of sandstone.

Examples of limestone arches that have been created by the action of a watercourse can be found on Cedar Creek in Virginia, US (Natural Arch), and on the Ardèche River in the south of France (Pont d'Arc).

WONDERS OF THE UNDERWORLD

Some of the most astounding geological wonders are to be found deep underground—caves and tunnels that range from individual, sculpted caverns to vast cave systems extending through literally hundreds of miles of vast chambers and connecting passageways. Like coastal arches, some of the most impressive are to be found in areas of limestone.

Limestone also has the important quality of being soluble as it is composed mainly of calcium carbonate, which dissolves in weakly acidic groundwater to form calcium bicarbonate, which is then carried away by the water. Because it is a sedimentary rock and is laid down in layers, limestone has horizontal "bedding planes" running through it. It also has vertical joints, and when water enters the rock through these joints it percolates down and then travels horizontally along the planes. Over

Nothing could be more lonely and nothing more beautiful than the view at nightfall across the prairies to these huge hill masses, when the lengthening shadows had at last merged into one and the faint after-glow of the red sunset filled the west.

Theodore Roosevelt

thousands of years, as the water makes its way through the rock, it dissolves the rock to form vertical sinkholes, spacious chambers, and horizontal tunnels at various levels throughout the limestone. This kind of erosion can be even more rapid in limestone that is below the water table and is constantly subject to water flowing through it, forming huge caves that are revealed when the water level drops.

Limestone is an extremely common type of rock, and major cave systems have been explored in many parts of the world. The world's longest known cave system is Mammoth Cave, in Kentucky's Mammoth Cave National Park (now a World Heritage Site), with more than 365 miles (580km) of tunnels already explored, and undoubtedly more to be discovered. Some of this extensive labyrinth is open to the public, as are caves in other parts of Kentucky, which has vast deposits of limestone.

The largest single chamber is to be found in Southeast Asia. The Sarawak Chamber, in Malaysia, is some 2,300 feet long and 1,300 feet wide, with a ceiling 230 feet high (700 x 400 x 70m), and was thought to be the largest enclosed space in the world until the discovery, in 2009, of the Son Doong Cave in Vietnam. The course of an underground river, its largest chamber reaches 650 feet (200m) in height, and is 330 feet (100m) wide and three miles (5km) long.

MINERAL TREASURES

Many of the world's limestone caves are renowned for the beauty of the mineral formations within them—their stalactites, stalagmites, and crystals. The words "stalactite" and "stalagmite" come from a Greek word meaning "to drip," and this is indeed how they are formed. As the eroding water, now rich in calcium bicarbonate, makes its way

gently down through the caves, it drips from ceilings and overhanging ledges and lands on cave floors. As the water evaporates, some of the calcium bicarbonate reverts to calcium carbonate in the form of calcite and, molecule by molecule, this builds up as hanging cones extending downward (stalactites) and others growing upward from the ground below (stalagmites). The process is a slow one, adding only a few inches (centimeters) every century, but these "dripstones" can grow to huge sizes, and even in large caves the upper and lower cones can eventually reach other and grow together to form a single pillar. These mineral art forms are widespread in limestone formations, but especially dramatic displays can be seen in the Lechuguilla Cave in the Carlsbad Caverns of New Mexico, in the Timpanogos Caves in Utah, and in Australia, including at the Jenolan Caves in the Blue Mountains of New South Wales.

Caves also provide conditions in which crystals of various minerals can form. For example, Wind Cave in South Dakota and Cody Caves in British Columbia, Canada, contain rare formations of calcite known as "boxwork." Thin, intersecting fins of the mineral project several inches (centimeters) from the cave walls, creating a honeycomb effect. Calcite can also form coral-like structures, and groups of long needles that form ice-like "frost flowers." Gypsum (calcium sulfate) can produce fine crystals known as "cave hair," and cave flowers with curved petals, but no crystals yet found compare with the giant gypsum crystals found in a mine near Naica in Chihuahua, Mexico. In what has become known as the Cueva de los Cristales are the largest natural crystals yet to be discovered—huge translucent beams of a form of gypsum called selenite. Some of these crystals are 36 feet (11 meters) long and weigh an estimated 55 tons!

Ancient Remains

Like many of the geological features that we have already looked at,
these wonders can be millions of years old, but if the word "fossil"
conjures up a boring and dry-as-dust image, think again. Fossils are
the only record we have of life on this planet in the distant past,
and without them our understanding of the world and our place in
it would be infinitely poorer. We should be very thankful for the
process of fossilization, and the wonders that it preserves.

When living organisms die—be they bacteria, plants, fungi, or animals—their physical remains are soon broken down, which is a good thing, or we'd be living on top of the bodies of everything that has ever lived and all the planet's nutrients would be unavailable to us. The agents that break the material down include bacteria that require oxygen in order to function, detritivores (such as slugs and insects), scavengers (such as birds) that consume and digest the soft tissues, and the forces of the weather that dissolve and break apart the remains. This process of decomposition returns the constituent chemicals to the atmosphere and the land, where they can be taken up by other living things. But what happens when, by chance, a dead organism manages to avoid the decomposers?

Well, there's a very slim chance that it may become a fossil.

Let's imagine a small dinosaur that has died beside a lake. While the body is still perfectly intact, heavy rains wash the body into the lake, where it sinks to the bottom and settles into the mud. The animal's soft tissue gradually decomposes, but the silt protects it from being torn apart by fish and crabs, and its skeleton remains intact. As the blood cells, proteins, and fats break down, the skeleton is eventually reduced to just the mineral elements (mainly calcium). Over thousands of years, sediment from the water settles to the lake bed, burying the bones under layer upon layer of mineral-rich sediment, and these minerals make their way into the bone, filling the spaces left by the departed organic matter and

When we hear the term "fossil" we usually think of bones, but the desert floor of the Petrified Forest National Park in Arizona is littered with huge chunks of fossilized wood composed of semiprecious minerals. On closer inspection, the fossilized remains of insects that lived hundreds of millions of years ago are visible trapped inside the wood.

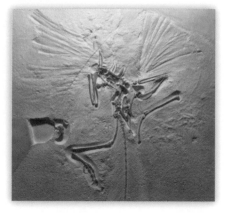

effectively turning the bone into rock. This is one of the ways in which a fossil can be formed, and it is known as "permineralization." Over millions of years, the layers of silt become compressed by the accumulation of further layers above them, and they become sedimentary rock surrounding the skeleton. The area may be uplifted and tilted by movements of the Earth's crust, wind and rain may erode the rock, and there is a very slight chance that eventually the stone-like skeleton of the dinosaur, complete with teeth and claws, may be found, dug up, and studied to add to our understanding of earlier life forms.

Fossils can also form when an organism leaves an impression in soft mud before it decomposes. When the mud hardens it creates a mold of the organism, and subsequent deposits of sediment may fill this and harden to form a cast. Fossilized footprints can also be formed in this way.

WHERE ARE FOSSILS FOUND?

As you may have guessed from the description above, the majority of fossils are found in sedimentary rock formations, and many of the locations that are famous for their eroded sandstone are also rich in fossils. In North America

these include parts of Texas, Arizona, Colorado, Wyoming, and the Dakotas, as well as several locations on the Eastern Seaboard where impressive trackways of dinosaur footprints have been found and bones have been discovered in the course of sandstone quarrying.

The sculpted canyons of Alberta's Badlands are rich in fossils, and the remains of more than 500 different species of plants and animals, including 40 species of dinosaur, have been unearthed in Dinosaur Provincial Park, to the southeast of Calgary. Today's barren moonscape of sandstone gullies, buttes, and hoodoos is a far cry from the landscape in which these giants thrived some 75 million years ago. Then it was a fertile river flood plain, rich in vegetation, and there are many fossils of freshwater vertebrates, from sharks and frogs to turtles and crocodiles.

Australia is rich in sedimentary rock and has many remarkable fossil sites, including Riversleigh in northwest Queensland, where the fossils of giant crocodiles, huge birds, and carnivorous kangaroos have been found. Australia is also home to fossils of the world's oldest recorded life forms. Stromatolites, structures formed by simple micro-organisms, discovered in the Pilbara region of Western Australia, may be as much as 3.2 billion years old.

In Europe, many of the finest fossils have come from limestone formations, and have been discovered when the rock was being quarried. Limestone from Soln-

Found near Solnhofen, Germany, in deposits of lithographic limestone, this fossil of Archaeopteryx displays features of both dinosaurs and birds.

hofen in Bavaria, Germany, has an unusu-
ally fine texture and, combined with the
fact that fossils formed here in conditions
of high salinity and in the absence of many
of the organisms that cause the soft tissues
to decompose, the rock contains extremely
detailed fossils. These include 145-million-
year-old examples of the earliest known
bird, *Archeopteryx*, in which the fine structure of the feathers can be seen, as well as
dragonflies and pterosaurs.

PETRIFIED FORESTS

Although the process is essentially the same
as the permineralization of bones, the fossilization of wood can be exceptionally spectacular. Petrification, or "turning to stone,"
occurs when uprooted trees are washed
away and become buried in mineral-rich
water and silt where there is very little
oxygen. Before the wood can rot, minerals permeate the structure of the wood and
replace the organic material. The process
can take less than 100 years, and the result
is effectively a stone tree. A range of minerals can be involved, but the most common
is quartz, a silicon-based compound. Quartz
is colorless, but it can crystallize in several
ways to form various semi-precious stones,
such as agate, and with different chemical
impurities, producing a range of colors. Petrified wood can also be composed of opal
(another silicon-based compound) and calcium carbonate (calcite). All of these miner-

als are extremely hard, and petrified wood
can be cut and polished to make jewelry and
other decorative items. The oldest petrified
wood comes from tree ferns and conifers,
but fossilized palms and hardwoods such as
elm and oak are also found.

Like other fossils, petrified wood tends
to occur in regions of sedimentary rock,
but especially where layers of volcanic ash
have been laid down, as this is a rich source
of silicon. In Yellowstone National Park,
an area of considerable volcanic activity,
there are many hundreds of petrified trees
thought to be some 30 to 35 million years
old. Some of them are standing upright, and
one is almost 50 feet (15m) tall.

Petrified forests are found in several
parts of the U.S., including New York State,
which has what may be the oldest petrified
trees in the world, some 360 million years
old. The country's largest concentration of
fossilized wood is to be found in the Painted
Desert of Arizona, in the Petrified Forest
National Park, where the 200-million-year-
old trees are almost solid quartz. This rock is
extremely brittle, and where movements of
the Earth's crust have caused the giant logs
to break, they have snapped so cleanly that
they look as though they were sawn. The
forms of quartz found here include amethyst
and citrine, which create scintillating patches
of purple and yellow in the crystal logs.

Quite a forest of petrified trees . . . They are converted into
beautiful specimens of variegated jasper . . . Fragments are
strewn over the surface for miles.

Lt. Amiel Whipple, leader of a scientific expedition in 1853

Rising and Falling

The eruption of a volcano or the sudden release of energy
when tectonic plates under tension finally move past each
other can be major catastrophes, but there are other slower
and less dramatic geological processes going on all the time.
Some of these happen on a relatively local scale, but others are
responsible for the largest features on the planet's face.

One of these processes is the strange phenomenon of "bradyseism." The term itself comes from two Greek words meaning "slow movement," and the best known example is to be found on the southwest coast of Italy, near Naples, not far from one of the world's most dangerous volcanoes, Mount Vesuvius, which erupted in 79 CE and destroyed the Roman cities of Pompeii and Herculaneum, killing thousands of people.

The small town of Pozzuoli sits in an area known since Roman times as the Phlegrean Fields, a circular formation some six miles (10km) in diameter. The area is actually a *caldera*, a plug of rock that sank down into the center of a volcano after the magma chamber beneath had emptied itself in an eruption some 35,000 years ago. Unfortunately, although the volcano may be dormant, it is far from extinct, and the pres-sure of gases and magma in the chamber, which is about 2.8 miles (4.5km) under-ground, slowly increases and decreases over time. The effect of this is to periodically raise and lower the caldera, and marks made by marine organisms on Roman marble col-umns in the town show that Pozzuoli has been dunked to a depth of at least 23 feet (7m) below sea level in the last 2,000 years and then raised up again. The town has risen by more than five feet (1.7m) on two occasions in the last 50 years. The second rise was accompanied by a small earth-quake, reminding residents of the Roman belief that the entrance to the underworld lies in the Phlegrean Fields.

MOUNTAIN BUILDING

It's a little hard to believe, but the world's highest mountain range is also one of the youngest, and it owes its existence to the

Evening sunlight bathes the peaks of Mount Everest and Mount Nuptse in Nepal. The Himalayan mountain range is both the largest and one of the youngest in the world, formed as the Indo-Australian and Eurasian tectonic plates began to converge. The Himalayan range is rising at about 0.2 inch (5mm) per year.

unhurried, but unstoppable, movements of the tectonic plates that make up the Earth's crust. The Himalayas, stretching for almost 1,500 miles (2,400km) across Asia from Pakistan to Bhutan along India's northern border, simply didn't exist when dinosaurs ruled the Earth. There was still ocean between India and the Tibetan plateau when they became extinct, but the gap was closing as the Indian land mass moved inexorably northward, colliding with the Eurasian tectonic plate some 50 million years ago. In one of the most spectacular examples of mountain building, or "orogeny," the slow-motion collision caused the ocean floor to crumple upward, eventually pushing layers of sedimentary rock (complete with marine fossils) more than five miles (8km) into the air and creating this magnificent range of "fold mountains." The process is still happening today as the Indian subcontinent continues to move northward at the rate of 2.6 inches (6.7cm) a year, sliding beneath the Tibetan plateau and forcing the Himalayan mountain range to rise about an inch every five years (5mm per year).

Movements of the Earth's tectonic plates have led to the formation of the majority of mountain ranges, from the Andes (where the Nazca and Antarctic plates are sliding beneath the South American Plate) and the Rocky Mountains, to the European Alps (formed by the African plate moving northward beneath the Eurasian plate) and the Eastern Highlands of Australia.

Once they are formed, mountains continue to be shaped by the forces of nature, such as erosion by glaciers and by rainfall. Tectonic uplift and the folding of the crust can cause once-horizontal layers to be tilted up almost vertically, and the erosion of softer layers leads to the saw-tooth appearance of the peaks of some mountain ranges. Hard, non-sedimentary rock can be eroded by the

The tilted and twisted sedimentary layers of this island in King Oscar Fjord off the east coast of Greenland reveal the folding of the Earth's crust.

Mountains are Earth's undecaying monuments.

Nathaniel Hawthorne

action of water freezing and expanding in cracks and crevices, breaking away sections of the rock and creating flat, almost vertical faces of the kind seen on the Matterhorn in the Alps on the Swiss/Italian border.

BOUNCING BACK

Glaciers are impressive, but they are nothing compared with the vast ice sheets that covered much of the northern hemisphere, and especially Canada, the northern USA, and northern Europe, in the depths of the last glacial period, some 18,000 years ago. In places the ice was more than one and a half miles (2.5km) thick, exerting unimaginable pressure on the land beneath it. As we have seen, the land is floating on semi-molten rock in Earth's mantle, the layer beneath the crust, and the effect of this pressure was to push the land down into the softer layers beneath, forcing the super-viscous material outward and causing the land beyond the ice sheet to bulge up slightly. As the ice melted and the weight was removed, the mantle material began to flow back beneath the depressed areas, causing them to rise up again while the raised areas sank back. Today, 12,000 years later, this gradual readjustment—which is known as "isostatic post-glacial rebound"—is still happening, although much more slowly, and it may continue for another 10,000 years.

The rate at which the land rebounds depends on the previous weight of ice and on the viscosity of the layers below, but it can be in the order of 0.4 inches (1cm) per year and can be detected by GPS surveying. The Great Lakes, on the border between Canada and the U.S., lie at the southern edge of the former ice sheet, so the northern shores of the lakes are rising while the southern shores are sinking, having a

significant effect on the water levels. Great Britain, too, is tilting, as Scotland rises and southern England sinks, raising concerns over flooding in the not too distant future. In Hudson Bay, in northern Canada, where the ice sheet was thickest, a series of successive shorelines shows the land has risen by more than 930 feet (300m) in the last 8,000 years, and it may have another 330 feet (100m) to go. The seashore has already retreated by more than 100 miles (160km) and Hudson Bay continues to shrink.

WHEN MOUNTAINS WALK

There are, as we have seen, several forces raising the land in different ways, but there is one force that is always working in the opposite direction—gravity. Its effects are ever present on all sloping surfaces, and are seen in the gradual erosion of rock and soil by glaciers and by water, but sometimes gravity produces dramatic and even catastrophic movement. Earth, mud, rock, and debris can suddenly be set in motion, triggered by an earthquake, a volcanic eruption, heavy rain or flooding, or by human activities, producing a landslide. This can range from a small-scale slide to the collapse of an entire hillside. Rockslides are part of the normal erosion process, and many mountain ranges show evidence of repeated rockslides, with cone-shaped piles of debris, know as "talus slopes," at their bases. Mudslides occur when a hillside becomes too weak to support its own weight, usually because it has become saturated with water. During periods of intense rainfall, deforestation increases the likelihood of this kind of landslide, as trees help to bind the soil together.

aquatic

phenomena

Having visited the wonders of the heavens, the atmosphere, and the land, we come finally to the waters of our planet—the inexorable movement of the tides and ocean currents, the awesome power of the waves, the impact of the microscopic marine life, and the beauty and energy of the vast polar ice sheets.

Tides

The waters of the world's oceans are forever on the move, pushed and pulled by the movements of the planet and the forces of wind and weather, channeled and deflected by continents and islands, rolled and tumbled by the topography of the seabed, but behind this seeming chaos there is an underlying order. The ebb and flow, rise and fall, of the tides is as regular as . . . the cycles of the Moon.

On coasts around the globe, approximately twice a day, the sea advances up the beach and then retreats again in a way that is so mathematically certain that the exact times and heights of high and low tide can be predicted years ahead for any location. Although each tide may look like a relatively small event—just a few more feet of water in the harbor—the fact is that countless trillions of tons of ocean are rising and falling, so what is the source of the regularity and of the power that can cause this to happen? The answer lies in the orbiting of the Moon around Earth, and of the Earth around the Sun.

We are generally aware that Earth exerts a gravitational pull on the Moon, but the fact is that the two bodies exert gravity on each other, although the influence of the Moon is far less than that of Earth because it has far less mass (less than one eightieth of Earth's). The Moon's gravity actually distorts the shape of Earth very slightly, but it has a far greater effect on the world's oceans, pulling them toward itself and creating a bulge of water on the side of Earth nearest to it. As the Moon orbits Earth, this produces a high tide approximately once every 25 hours. The Moon is also pulling Earth toward it slightly, but as gravity decreases with distance, the Moon has less influence on the ocean on the opposite side of the globe. It therefore pulls Earth away from the ocean and this creates another bulge on the side farthest from the Moon. This is why there is a high tide every 12.5 hours.

Drawn by the pull of the Sun and the Moon, the sea retreats from the shoreline and then advances again, unstoppable and as regular as clockwork. The influence of the tides, and of the gravitational pull of the Sun and Moon to which the tides respond, is evident in the "biological rhythms" of organisms that live between the high and low water lines.

SPRINGS AND NEAPS

That all seems logical enough, but why then doesn't the tide always rise to the same the height? This is where the influence of the Sun comes into play. The Sun, too, exerts gravity on Earth (otherwise our planet wouldn't be orbiting the Sun), and it, too, pulls the water toward it. Although the Sun is vastly more massive, it is much farther away, so its pull on the water is less than half that of the Moon, but it effectively creates its own tides on a 24-hour cycle and these interact with the Moon's tides. The extent to which the tides rise and fall therefore depends on the

relative positions of the Sun, Earth, and the Moon. As we have already seen, those relative positions also determine the phases of the Moon, and (surprise, surprise) the heights of the tides follow the same cycle as the Moon's phases. When the three heavenly bodies are in line with each other (that is, when we see a full Moon or a new Moon), the influences of the Sun and the Moon are combined and the tides reach their most extreme high and low points (especially when the Moon is at the nearest point of its orbit to Earth and the Earth is at its nearest to the Sun). These are called "spring tides."

When the Sun and the Moon form a right angle with the Earth, and we see the Moon in its first quarter or third quarter, the influences of the Sun and Moon are acting at 90° to each other and the rise and fall of the tides are at their minimum. These are called neap tides.

 These two photographs of a harbor in the Bay of Fundy demonstrate the dramatic tidal range of this region.

VARIATIONS ON A THEME

Of course, a host of factors affects the actual timing and the height of the tide in any particular location. The water doesn't only move vertically. As well as the rise and fall there is the horizontal motion that occurs as the water pushes into channels and onto shelving shores. This is the tide's ebb and flow. An offshore island, for example, can cause the rising water to reach the coast from two different directions with a delay between them, creating a double tide locally. The movement of the water can be influenced by the depth and by the contours of the seabed, and even by the salinity of the water, which affects its density. Importantly, the sea will rise higher when funneled into a narrow bay than it will on a long, straight shore, and there are some remarkable examples of this.

The best-known is the Bay of Fundy, a 170-mile- (270km-) long inlet to the east of Maine, between the Canadian provinces of New Brunswick and Nova Scotia. Twice a day, more than 100 billion tons of water—more than the world's rivers discharge in a day—flow into the bay, and the tapering shape of the inlet funnels the water into a steadily narrower space. Heightening the effect, the period between low and high tide is the same as that taken by a wave to travel the length of the inlet, and this is believed to produce a form of resonance that raises the water level even higher. At the time of the most extreme spring tides, the difference between high and low water in parts of the Bay can be as much as 55 feet (17 meters), producing the world's greatest tidal range. (Ungava Bay in northern Quebec, which has a long-standing rivalry with the Bay of Fundy, has now been shown to have the greatest *average* tidal range.)

WHIRLPOOLS

Long the stuff of legends, whirlpools that can pull a boat into the depths really do exist. In several parts of the world, including the Bay of Fundy, strong tidal flows combine with contoured sea beds and jutting coastal headlands to produce fierce currents that can form a swirling vortex. A whirlpool that forms at certain stages of the tide between Deer Island and the coast of Maine, at the mouth of the Bay of Fundy, has been named the Old Sow, possibly because of the loud noise made by the churning water. Small boats have frequently been pulled into the rotating current there, and though most have been able to escape, lives have been lost.

The world's most powerful whirlpool forms in a narrow straight called Saltstraumen on the coast of Norway. Here, each tide forces some 400 million tons of water through a channel less than 500 feet (150m) wide, and the resulting *maelstrom* (from the Dutch, meaning "grinding stream") produces vortices up to 33 feet (10m) across and 16 feet (5m) deep.

Other whirlpools are found farther north on the Norwegian coast, in the Naruto Strait in Japan, on the Sunshine Coast of British Columbia, Canada, and in the Gulf of Corryvreckan off the coast of Scotland.

TIDAL BORES

The high tides can be thought of as two waves forever traveling around the globe, each with a wavelength (at the equator) of half of Earth's circumference—about 12,000 miles (20,000km). When one of these waves reaches the mouth of a river, it can hold back the flow of the river, or even cause water to flow back upstream, but once in a while, in certain locations, it can do much more than this. On the highest spring tides, which generally occur around the time of the spring and fall equinoxes (late March and late September), when the Moon is above the equator, rivers can experience a tidal bore, a wave that actually travels up the river, sometimes for many miles. This is a true "tidal wave," a term that should not be applied to a tsunami (which is not tide-generated). Tidal bores occur in many rivers around the world, but generally in those that normally have a large tidal range. These include the Ganges (India), the Daly River (Australia), the River Dee (Scotland), and, formerly, the Seine in France. Dredging and management of the lower reaches of the river put an end to this bore in the 1960s. Some rivers, however, have achieved fame not only for the size of their bores but also for their attraction as surfing destinations.

The River Severn in England has a tidal range almost the equal of the Bay of Fundy, and at the time of the highest spring tides the funneling effect of the wide estuary sends a wave several miles up the normally placid river against the current. For the last 50 years the event has attracted dozens of surfers keen to ride one of the world's longest waves. The official long-distance surfing record of 7.8 miles (12.5 km) was set by Brazilian surfer, Picuruta Salazar, riding a wave known as the "Pororoca" on the Amazon River. This tidal bore, which can reach a height of 13 feet (4m), has reportedly been detected 124 miles (200km) upriver from the estuary. In recent years, the world's largest bore, reaching a height of 30 feet (9m) and known locally as the Black Dragon, has attracted American and European surfers to the city of Hangzhou on the Qiantang River in China, where an unofficial distance record of 9 miles (14.4km) was set by another Brazilian surfer, Sergio Laus, in 2009.

As we have seen, the Bay of Fundy has a rare combination of geographical and tidal features, so it comes as no surprise to learn that the Shubenacadie River, which opens into the farthest tip of the Bay, experiences a tidal bore with every high tide. The height of the wave can be as little as one foot (30cm), but as much as 10 feet (3m) on the highest springs.

There is a tide in the affairs of men

Which, taken at the flood, leads on to fortune;

Omitted, all the voyage of their life

Is bound in shallows and in miseries.

William Shakespeare, *Julius Caesar* (Act IV, Scene ii)

TIDES

What they are:

Tides are the rise and fall of the sea that occurs approximately twice daily at all points around the Earth.

What happens:

The gravitational pull of the Moon draws water toward it, causing a rise in sea level on the Moon-ward side of the Earth. The Moon also attracts the Earth more than that it does the water on the far side of the planet, and this results in a corresponding bulge on the side of Earth farthest from the Moon. As Earth spins and the Moon changes its relative position, these two bulges in the world's oceans move around Earth, creating two tides a day. The Sun's gravitational force also pulls water toward it, and when the Sun and Moon are in line with Earth these influences reinforce each other, producing higher high tides and lower low tides.

These are call spring tides. When the two gravitational forces are pulling at right angles to each other there is less variation between the high and low tides. These are called neap tides.

Where to see them:

Tidal rise and fall can be seen on all ocean coasts, although the tidal variation is more dramatic in certain parts of the world where the geography and seabed contours channel the rising water into inlets.

When to see them:

The tidal ebb and flow is continuous and can always be seen in action, but because the tides are of such vital importance for commercial shipping, fishing, and recreational boating, tide tables are available to tell you the exact state of the tide at any moment in any location.

The upper of the two diagrams below shows the position of the Earth and Moon at Spring tides, while the lower illustrates their positions at Neap tides.

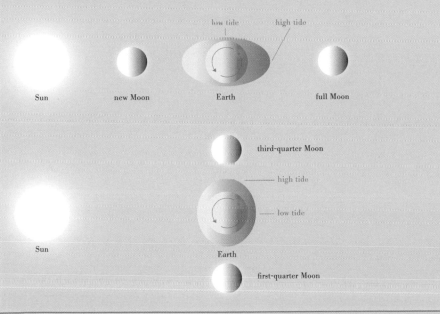

low tide   high tide

Sun   new Moon   Earth   full Moon

third-quarter Moon

high tide

low tide

Sun

Earth

first-quarter Moon

Oceans in Motion

Although their rise and fall pushes and pulls vast quantities of water through straits and channels, up rivers, and onto beaches, the tides are effectively a vertical movement of the water; but the world's oceans also contain currents that flow in fairly consistent directions around the globe. Carrying heat energy and nutrients, these currents have a major effect on the climate around the globe, and they determine the locations of the world's great fisheries.

If you thought the oceans were just vast lakes of stationary water with a little surface disturbance in the form of waves, then think again. These waters are on the move, and on a grand scale, with separate surface currents circulating in the northern and southern hemispheres while deeper currents move in different, and even opposite, directions beneath them. Both the surface currents and the deep currents are driven, in different ways, by heat energy from the Sun and by the Earth's rotation on its axis.

SURFACE CURRENTS

The existence of large-scale and predictable surface currents in the world's oceans has been known for hundreds of years, as they were of vital importance to sailing ships. Taking advantage of the currents could knock weeks off a long sea voyage, and battling against them could make it impossible for a ship to reach its destination. Now, with the technology that is available to oceanographers, the nature of these currents and the mechanisms behind them are understood in considerable detail.

The initial impetus for the currents comes from the Sun, which, through the differential heating of areas of the planet's surface, causes the winds to blow. As the winds generated by areas of high pressure in the subtropics pass over the ocean, they drag the surface of the water with them, and the effect of this drag can extend some 330 feet (100m) beneath the surface. As the water flows away from the equator—northward in the northern hemisphere and southward in the southern—another factor comes into play, affecting the course of the flow: the Coriolis Effect.

Foul weather and turbulent seas, as depicted in this 1680 painting by the Dutch artist Willem van de Velde, were not the only problems facing seafarers in the days of sail. An understanding of the ocean currents was vital to making a speedy passage and any miscalculation of their effect would drag a ship dangerously off course.

THE CORIOLIS EFFECT

Earth is constantly spinning eastward, or counter-clockwise, when looked at from above the North Pole, and the world's oceans are moving with it. Earth's circumference is greatest at the equator, and therefore the eastward velocity of Earth's surface, and the ocean, is greater at the equator than it is farther north or farther south. At the poles the velocity is zero. Imagine, then, a volume of water flowing northward from the equator. As it moves north it maintains its constant eastward momentum, but the Earth beneath it is moving more slowly, so the water moves eastward, or to the right, in relation to the Earth. This is the Coriolis Effect, and it increases as one moves farther away from the equator. The effect is to cause currents of water (and of air, for that matter) to turn to the right as they move north from the equator and to the left as they head south from the equator. The farther from the equator the currents travel, the more they bend toward the east, and the result is a huge circular flow around each of the world's major expanses of ocean— the North Atlantic, South Atlantic, North Pacific, South Pacific, and Indian Ocean. These rotating systems, or "gyres," rotate clockwise in the northern hemisphere and counter-clockwise in the southern hemisphere, and they constitute the major surface currents, together with the Antarctic Circumpolar Current that flows from west to east around Antarctica.

WARM OCEAN CURRENTS

Typically, the current on the western side of a gyre, which flows toward the pole, is narrow and composed of warmer water that has been heated by the Sun in the equatorial regions, while the eastern, equator-bound current is much broader and cooler. In the North Atlantic, the "western boundary current" is the Gulf Stream, flowing up the east coast of the US from the Gulf of Mexico to North Carolina and past Newfoundland. As it flows northward the current intensifies. It is estimated that the Gulf Stream—which is about 60 miles (100km) wide and 3,300 feet (1km) deep—flows through the Florida Straits at the rate of about 30 million tons of water per second. By the time it approaches Newfoundland the rate reaches 150 million tons of water per second.

Off the Canadian coast the current heads across the Atlantic to Europe and becomes the North Atlantic Drift. The temperature of the Gulf Stream is considerably higher than that of the surrounding ocean, and when this water reaches the eastern Atlantic it flows northward past the British Isles and Scandinavia. The warm water and the moisture-laden air currents that are created by it have a significant impact on the climate of northwest Europe. This maritime

Roll on, deep and dark blue ocean, roll. Ten thousand fleets sweep over thee in vain. Man marks the earth with ruin, but his control stops with the shore.

Lord Byron

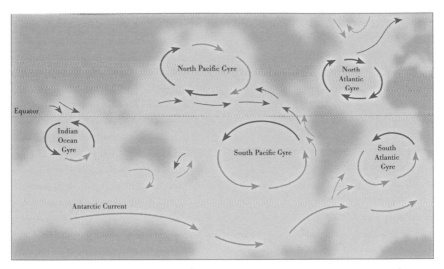

The world's ocean gyres rotate clockwise in the northern hemisphere, and counter-clockwise in the southern hemisphere, carrying warm water away from the equator.

influence can be seen in the fact that palm trees can grow in the southwest of England and even in the Hebridean Islands off the west coast of Scotland, which are farther north than Polar Bear Provincial Park in northern Ontario (where there are very few palm trees!). A similar situation exists in the Northern Pacific, where the Kuroshio Current brings warmer water northward past Japan and the North Pacific Current carries this to the west coast of Canada and the south coast of Alaska, bringing a mild climate and supporting the growth of temperate rainforest in these regions.

In the southern hemisphere, the warm south-flowing currents are the Brazil Current in the South Atlantic, which warms the east coast of South America, the Agulhas Current that flows down the east coast of Africa, and the East Australia Current (the western part of the South Pacific Gyre) that affects the climates of eastern Australia and New Zealand.

COLD CURRENTS AND UPWELLINGS

The winds that circulate around the subtropical highs also push currents of cooler water toward the equator on the eastern side of the gyre. These are typically slow, broad, and shallow. In the Atlantic, the "eastern boundary current" is known as the Canary Current, after the Canary Islands through which it flows. As the wind blows down the northwest coast of Africa, the surface current flows to the right, away from the shore, as a result of the Coriolis effect, and this causes water to be drawn up from the depths to replace it. This is known as an "upwelling," and it is a feature of eastern boundary currents and the winds that power them, affecting the western coasts of continents around the world. Upwellings bring nutrients up to the ocean's surface layer, fueling the growth of phytoplankton, which in turn are eaten by zooplankton, and so on up the food chain. This leads to a rich and diverse ecosystem and highly productive fisheries. Upwellings occupy about one twentieth

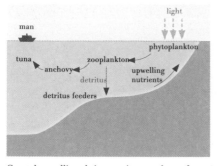

Coastal upwellings bring nutrients to the surface.

of the ocean surface, but they account for about one fifth of the marine harvest.

In the northern Pacific, the equivalent current is the California Current, which brings cooler water down the west coast of North America and is further cooled by the upwellings caused by the southward-blowing winds. This cool water produces the familiar fogs of the Californian coast, but the nutrient-rich waters also support highly productive fisheries.

The cool northward-flowing currents of the southern hemisphere are the West Australia Current on the eastern side of the Indian Ocean, the Benguela current up the eastern side of the South Atlantic along the coast of Africa, and the Humboldt current that flows up the Pacific coast of South America, past Chile, Peru, and Ecuador. The upwelling system of the Humboldt is one of the most productive marine ecosystems on Earth.

 The El Niño climatic event can bring extremely heavy rainfall, flooding and severe mudslides to countries on the west coast of South America.

The cold Benguela and Humboldt currents cool the air above them, which is then drawn onshore by the rising currents of warm air on the land, but the cold moisture-laden air remains trapped below warmer air and cannot rise. Therefore clouds cannot form and there is no precipitation, so these currents produce arid conditions on the land, giving rise to the Namib Desert in Africa and the Atacama Desert of Chile, both of which are cold and dry. The currents do, however, bring dense and persistent fog over the water and, in the case of the Namib Desert, up to 50 miles (80km) inland. The coast of northern Namibia is known as the Skeleton Coast, both because of the hundreds of fog-bound ships that have been wrecked on its rocky shores and for the many whale skeletons along its beaches.

The strongest of the world's currents, the Antarctic Circumpolar Current, circles Antarctica at the rate of about 2.5 miles per hour (4km/h), shielding the polar region from the warmer waters farther north and helping to keep down the temperatures on the frozen continent.

EL NIÑO

As we have seen, the Humboldt current and its upwellings maintain cool water along the west coast of South America, providing highly productive ocean fisheries and relatively dry atmospheric conditions over the land. At the same time, the warm, south-flowing current in the western Pacific brings copious rainfall to regions such as Indonesia and Queensland in northeast Australia. However, once in a while the situation gets turned on its head, with serious consequences for the climate over a large proportion of the globe.

The phenomenon is called El Niño, which is Spanish for "the little boy" and refers to the infant Jesus. It takes its name from a small-scale local change in ocean temperature that occurs around Christmas time. The sequence of events is complex, and not completely understood, but broadly speaking the trade winds blowing up the South American coast from the southeast weaken for an extended period and the temperature of the waters off the coast of South America begins to rise, indicating the arrival of El Niño, which happens once every three to seven years. The atmospheric pressure in the eastern Pacific decreases, while it rises in the western Pacific, pushing air and warm water across the Pacific from west to east, where it replaces the cold water of the Humboldt current.

The effect of this is to bring powerful thunderstorms and vastly increased rainfall, with the potential for flooding, to the western part of South America and as far inland as southern Brazil and northern Argentina. The normally bountiful fishery off the coast of Peru, deprived of the cold-water upwelling and the nutrients that it brings, collapses, sometimes catastrophically. At the same time, countries in the western Pacific and around the Indian Ocean experience lower ocean temperatures and suffer exceptionally dry years with the risk of drought and famine.

On a broader scale, the belt of warmer water that extends across the eastern Pacific affects the pattern of low- and high-pressure areas in the North Pacific. Changes in the major air currents increase precipitation in northern Mexico and the southwest of the U.S., while bringing milder winters and less snow to the Midwest, the Northwest, and much of Canada. Indeed, the El Niño event in effect threatened to jeopardize the 2010 Vancouver Winter Olympics.

El Niño is sometimes followed by an extreme swing past a return to normal, with a deep cooling of the east Pacific and an intensification of the trade winds. This reversal is referred to as La Niña—"the little girl"—and the full scenario is known as the El Niño Southern Oscillation (ENSO).

OCEAN EDDIES

We have seen that narrow tidal currents can give rise to whirlpools many yards (meters) in diameter, so you can just imagine the kind of swirls that an ocean current can create. They may not be visible with the naked eye, but the images revealed by satellite imagery are truly impressive, as eddies ranging from 10 miles (16km) up to 300 miles (500km) across are spawned from the major currents, and especially by the Gulf Stream. These ocean eddies occur when the current starts to veer slightly from side to side, eventually forming a large meander that finally becomes nipped off and circulates independently. Warm-water eddies rotate clockwise in the northern hemisphere because of the Coriolis effect, while cold water eddies rotate counter-clockwise. The opposite is true in the southern hemisphere.

These giant swirls—known as "mesoscale eddies"—are tracked using satellite observation because they are large enough to affect commercial shipping and naval operations. They tend to differ from the surrounding water not only in temperature, carrying heat around the oceans, but also in salinity. The largest of them can last for months or even years.

GREAT PACIFIC GARBAGE PATCH

If you have ever watched a floating leaf stuck in an eddy at the edge of a river, you may be able to predict what happens at the center of an ocean gyre. The circular current creates an area of stable, slow-moving water in the middle, and any material that is drawn into this area can stay there for a very long time. Flotsam such as driftwood and seaweed will eventually decay, disintegrate, and dissolve, to be taken up by other living things, but sadly not everything that finds its way into the ocean is part of the natural cycle.

In the central region of each of the world's major ocean gyres there is now an unbelievably huge accumulation of plastic debris made up of bags, toys, fishing nets and floats, household items, flip-flops, crates, packaging, billions of bottles . . . the list is endless.

At the center of the North Pacific Gyre, between Hawaii and California, a region of the ocean that has been dubbed the Great Pacific Garbage Patch covers an area at least the size of Texas. Some estimates put it at twenty times that size. The reason there is doubt about the extent of the garbage patch is that the majority of the debris is not floating. None of this plastic is biodegradable—it will never be broken down into its elements or compounds simple enough to be used by living things—but it does break down into ever smaller pieces. These tiny fragments, known as "nurdles," are suspended in the top 100 feet (30m) of the ocean and can only be detected by research vessels. Nurdles have the property of leaching toxins into the water, and absorbing toxins from it. As the plastics break down further, ultimately into individual molecules, they are ingested by living things that are then affected by the chemical compounds the plastics contain. (Some of these compounds emulate sex hormones and can disrupt the reproductive processes of the animals.) Taken up by the smallest creatures, such as krill, the plastics and the toxins make their way up the food chain, becoming ever more concentrated as they do so. And that food chain includes us.

An ocean eddy is illuminated by a buildup of light-blue phytoplankton southeast of Hokkaido, Japan. The concentration of such phytoplankton is often greatest at the boundaries of eddies.

The Ocean Waves

Regardless of tides and currents, the surface of the ocean and other large expanses of water is continuously on the move, rising and falling in waves that range from tiny ripples to gigantic rollers. In this section we look at how waves are formed, at the strange behavior they can impart to the water, and at the devastating effects they can have on ships and coastal structures, and even on the lives of people dwelling inland.

Water waves have several properties. As they travel across the surface of the water, energy is transferred in the direction of travel, but the individual water molecules do not make an overall horizontal movement. In fact they trace a circular path as they move back, up, forward, and down. A sequence of waves can be described in terms of its wavelength (the horizontal distance between successive crests or successive troughs), its amplitude (the vertical distance between the trough and the crest), and its frequency (the number of crests passing a given point in a given amount of time). The steepness of a wave is the ratio of its amplitude to its wavelength.

BORN OF THE WIND

Most waves are caused by the pressure of the wind blowing across the surface of the water. The equilibrium or rest position of the water surface is, ideally, flat. The pressure of the wind imparts energy to the water and creates a disturbance, causing waves to be propagated across the surface. Without the input of further energy, the waves gradually die down as various forces act to restore equilibrium.

At the smallest scale, the force pulling the water flat is capillary action, which we see as surface tension. Whereas water molecules in the body of the water are being pulled equally in all directions by their neighbors, molecules at the surface are only being pulled sideways and downward, which makes the surface behave like a taut skin. A very light breeze will gently touch the surface and dimple it with tiny capillary waves, or ripples, with a wavelength of just an inch or so (a few centimeters). These are

Driven by storm-force winds, ocean waves turn to surf as they approach the shallower waters off the coast of Oahu, Hawaii. Arguably the most popular of the Hawaiian islands among surfers, the waves off the coast of Oahu very occasionally reach heights in excess of 40 feet (12m).

direction of waves

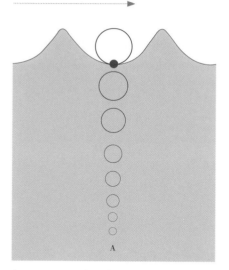

As a wave passes through the water, individual water molecules trace a circular path (A). The amount of movement decreases with depth.

known as "cats' paws," because their dark middles (where the light is being reflected away from us) and brighter margins give them the appearance of paw prints. When the breeze passes, the water quickly regains its mirror-like calm.

Once the wind is strong enough to create larger waves, it is gravity rather than capillary action that is acting to restore the equilibrium of the water surface. The size to which gravity waves grow depends on a number of factors: the speed of the wind; the size of the area over which it blows; and the time for which it blows. The higher the waves become, the more energy the wind can impart to them, and a strong wind blowing for a long time over a long distance (or "fetch") and across a wide front can produce huge "wind waves" that become a serious problem for shipping on the open ocean. The waves that continue after the wind has abated are known as a "swell."

As waves approach the shore and the water becomes shallower, they slow down and the wavelength is reduced as they catch up with each other. At the same time, even though they are losing energy through friction with the seabed, the amplitude of the waves (their height) increases and they become steeper. Eventually, when the water becomes sufficiently shallow, the steepness becomes too great and the waves topple forward and break. When one part of a wave reaches shallow water before the rest, by crossing a submerged reef, point, or sandbar, the break will move along the wave from one end to the other.

RIP CURRENTS

Every year, swimmers fall prey to the waves' final trick. Even when it has broken on the beach, a wave still has energy—the potential energy that comes from being above sea level. When waves break over a sandbar—even a submerged one—that then impedes their flow back to the sea, the water can flow rapidly out through a gap in the sandbar or in a direction parallel to the beach toward the end of the bar. Under these conditions, known as a "rip current," swimmers can suddenly find themselves being carried quickly out to sea, while non-swimmers standing in relatively shallow water can have their legs pulled out from under them and be swept into deeper water. Trying to fight against the current is fruitless, and the only way to get back is to swim out of the side of the rip current into calmer water.

ROGUE WAVES

For centuries mariners have spoken of monstrous towering waves that arrive out of nowhere in the open sea and sink large ships without trace, but such tales remained entirely

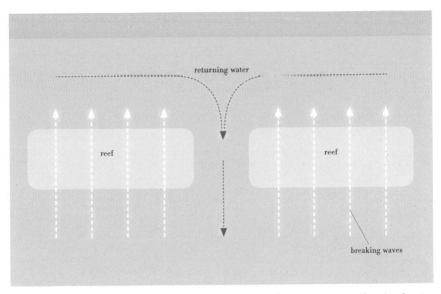

Rip currents occur when waves break over a reef or sand bar and the returning water is forced to flow back around the end of the bar or through a gap in it.

unsubstantiated until the 20th century, when the captains of several ships reported experiencing and surviving such waves. Then, in December 1978, the 850-foot- (260m-) container ship MS *München* sank in the North Atlantic with the loss of all the crew in a storm that, although it was extremely violent, should not have posed a hazard to this modern and well-designed vessel. One of the ship's lifeboats, which had been mounted 66 feet (20m) above the waterline, was recovered, and its mounting pins were found to be severely bent, indicating that it had been ripped off the ship by an enormous force. This added further evidence, albeit circumstantial, for the existence of freak waves, but on New Year's Day in 1995 came positive proof. The oil-drilling platform *Draupner*, located in the North Sea off the Norwegian coast, was hit by a giant wave, and this time there were measuring instruments recording it. In a sea with steady wave heights of about

40 feet (12m), this lone wave measured 84 feet (25.6m) from trough to crest.

Rogue waves are not only a fact—studies suggest they may also be fairly common. Their danger lies not just in their size, but in their incredible steepness. A large ship can ride over a giant swell, like a truck over a hill, but when it meets a vertical wall of water the pressure of the impact can be as high as 10 tons per square foot (100 tonnes per sq m), more than six times the pressure that ships are designed to withstand. The causes of rogue waves remain obscure, but explanations include the meeting of two opposing currents, the meeting of two wave trains that reinforce each other and combine their amplitudes, and the possibility—drawn from quantum physics—that a normal wave might occasionally draw energy from the waves in front of it and behind it to become a superwave.

TSUNAMIS

Tsunamis, gigantic waves that can cause terrible damage and loss of life when they strike coastlines, are often mistakenly referred to as "tidal waves." In fact they are not caused by the tides, nor are they generated by the wind, as the other kinds of waves we have been examining are. Their origins usually lie in earthquakes that involve sudden vertical movements of Earth's crust beneath the ocean, most commonly when the edge of one tectonic plate moves beneath another at a subduction zone, causing one part of the ocean floor to fall and the other to rise. The shock wave displaces a huge volume of

A towering crest dwarfs the fishermen and their boats in The Great Wave off Kanagawa, a painting by the Japanese artist Hokusai. The majority of tsunamis occur in the Pacific Ocean.

water and creates one or more waves that propagate out from the epicenter. The wave that travels out from the downward-moving side of the fault is preceded by a trough, while a crest leads on the upward-moving side. These waves are not high—often less than three feet (1m)—and they have a wavelength that can exceed 60 miles (100km), so as they travel across the deep waters of the ocean at speeds in excess of 500 miles per hour (800km/h) they go almost unnoticed by shipping. However, as they approach the coast and the water depth decreases, they behave in the same way as a wind-generated wave. They slow down dramatically, the wavelength decreases, and the amplitude steadily increases, with the water gradually rising into a giant wave.

On the side of the earthquake zone that fell, the first effect to reach the shore is a trough, rather than a peak, so people on the beaches may see the water draw back to a level even lower than that of a normal low tide. The tsunami itself is rising offshore and will strike within minutes, and people in the know are heading for higher ground as fast as they can. Those who can't read the signs, meanwhile, are watching in wonder or even walking down the newly exposed beach. On the side of the fault that has risen, coastal inhabitants don't receive even this much warning as the impending wave approaches.

When it reaches the land, traveling at about 50 miles per hour (80km/h), a tsunami can be as much as 100 feet (30m) tall and has enormous energy, but it is the sheer volume of water that is so destructive. A tsunami can rush up to 3,000 feet (1km) inland, carrying everything in its path—trees, soil, people, cars, and even buildings. As it retreats, it drags objects and people with it to the sea, and often the subsequent waves are even larger, doing yet more damage. The word *tsunami* means "harbor wave" in Japanese, and while boats in deeper water off the coast may experience only a large swell, those in harbor can be flung far inland or reduced to matchwood.

The most destructive tsunami in living memory hit the coast of many Asian countries on December 26, 2004, with a death toll of more than quarter of a million people. It was triggered by a 9.3-magnitude earthquake off the west coast of Sumatra in the Indian Ocean, the second most powerful ever recorded (the most powerful being the 9.5-magnitude quake that hit Valdivia, Chile, in 1960). The event prompted agencies worldwide to install tsunami monitoring equipment and set up early-warning systems in the hope of preventing such tragic loss of life in the future.

Major tsunamis occur, on average, once every 15 to 20 years. The vast majority of these occur in the Pacific, because

of the high level of seismic activity around the Pacific plate, but not all tsunamis are caused by earthquakes. A small number are caused by volcanic eruptions, meteor strikes, and massive coastal or underwater landslides (sometimes as the result of an earthquake) that displace large volumes of water. Landslides can create what are known as "megatsunamis," and the highest wave ever recorded was produced by one. In 1958, following an earthquake, a giant slab of rock with an estimated mass of 90 million tons fell into the sea at the head of Lituya Bay, a 9-mile- (14km-) long inlet on the coast of Alaska. The resultant wave, possibly amplified by reflection from the steep walls of the bay and by seiche wave motion (see p. 206), reached a height of more than 1,700 feet (520m), stripping the soil and trees from the surrounding area up to that height and destroying two fishing boats at the mouth of the inlet. Incredibly, one boat and its father-and-son crew survived after being hurled over a forested spit of land.

The size of the tsunami following the meteor impact in the ocean off the Yucatan Peninsula some 65 million years ago, which is thought to have caused the extinction of the dinosaurs, can hardly be imagined.

STORM SURGE

Sometimes referred to as "meteotsunami," storm surges are large waves driven by powerful hurricanes. As a tropical storm moves across the ocean, the pressure of the wind causes the water to build up in a wall in front of it. When the storm makes landfall, this bulge of water behaves like a tsunami and can cause extensive flooding and damage, especially when combined with high wind waves and heavy rainfall. When a storm surge strikes at the same time as a high tide, the destruction is even greater. The storm surge produced by Hurricane Katrina, which hit Louisiana in 2005, reached a height of 25 feet (7.6m)—one of the largest ever recorded in the U.S.

SEICHE

Atmospheric disturbances such as strong winds and downdrafts can have particularly strange effects within enclosed or semi-enclosed bodies of water, as a result of resonance. Every such body of water—be it a lake, inlet, bay, harbor, or even a swimming pool—has what is called its "eigenfrequencies" (meaning "own frequencies"). These are resonant frequencies at which it will naturally oscillate, and they are determined by its shape, length, and depth. If an undulation of the water at a matching frequency occurs, as the result of waves entering it or because of seismic or atmospheric activity, a standing wave ("sloshing") can be set up. Known as a *seiche*, this kind of wave simply moves up and down rather than progressing across the surface of the water, and it can be highly destructive. If a storm surge happens to initiate a seiche, for example, the degree of flooding and damage is greatly increased as the exaggerated wave moves up and down at the shoreline. Earthquakes often cause seiches in swimming pools, and even in lakes hundreds of miles away. Certain harbors in the Mediterranean experience seiches every few years as the result of rapid changes in atmospheric pressure. The Illinois coast of Lake Michigan frequently experiences small seiches, and several people drowned in Chicago in 1954 when a 10-foot (3m) seiche washed them off the rocks. Even bodies of water as large as the Baltic and Adriatic Seas experience seiches from time to time, as St. Petersburg and Venice know to their cost.

CLAPOTIS

Another form of resonance occurs when steep-sided boundaries, such as harbor walls, reflect waves back in such a way that the incoming and returning crests and troughs alternately reinforce each other and cancel each other out. The effect is to produce a standing wave—known as a *clapotis*—that rocks up and down with twice the amplitude of the original waves. Where this occurs, the action of the water can eat away at the bottom of the wall or cliff, causing severe erosion.

When the wave train hits the vertical surface at an angle, it can interact with the reflected waves to create cross-hatching on the surface of the water, an effect known by the French term *clapotis gaufré*, which translates as a "waffled clapotis."

MICROBAROMS

Standing waves oscillate up and down with a regular frequency, and they affect the air above the water surface in much the same way as the cone of a loudspeaker, sending out a very low-frequency sequence of compressions and rarefactions of the air, known as "microbaroms." These infrasound waves—typically with a frequency between 9 and 15 waves per minute—can be picked up by sensors many hundreds of miles away. Standing waves can also be generated by storms in the open ocean, and these too can be "heard" at enormous distances.

Inundated by the powerful storm surge, streets close to the Central Business District of New Orleans remain flooded in the aftermath of Hurricane Katrina.

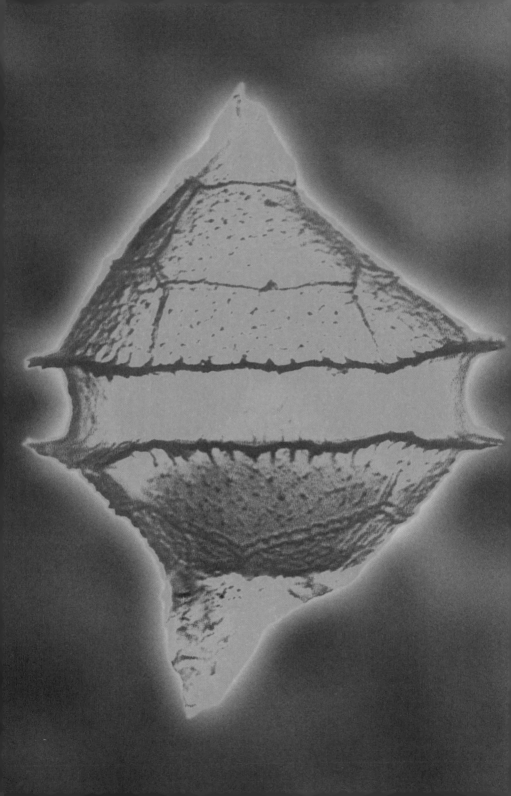

Biological Phenomena

So far we have been looking at the remarkable motions of the world's oceans—the currents, tides, and waves that manifest the energy that the seas possess. Now we turn to some of the wonders that living things bring to the blue two-thirds of the planet.

Two particularly strange and delightful phenomena owe their existence to the tiniest of life forms, single-celled organisms that have the bizarre ability to convert chemical energy into light—a quality known as "bioluminescence." There are a few land-based creatures, such as fireflies and glow-worms, that can do this, as well as certain fungi that lend an eerie light to the forest, but the seas have a wealth of glowing things.

For centuries, mariners have watched tiny sparkling, blue-green lights that decorate the bow wave and shimmer in the wake of boats and ships traveling through the night. Now we know that these lights are produced by members of a phylum of algae called dinoflagellates, a name that comes from the Greek for "whirling whip," as they have a whip-like appendage with which they propel themselves through the water. The reason that these microscopic plankton illuminate the water around boats is that they are responding to the movement, and it is thought that this is a defense mechanism. Dinoflagellates are right at the bottom of the food chain, and plenty of creatures, such as shrimps and small fish, feed on them. The theory goes that by emitting light the algae attract larger predators that prey on the would-be feeders, who quickly leave the scene to save their own skins, and research has shown that the larger predators, such as squid and fish, do indeed have more success when the lights are turned on.

MILKY SEAS

On a larger scale, the crews of sea-going vessels have long reported sailing, sometimes for hours on end, through night-time waters that gently glow with a pale blue light, giving the sea a milky appearance. Without an adequate explanation, the phenomenon remained in the realm of myth until the last 100 years, when hard data began to be gathered. Dinoflagellates were ruled out as the origin because the light was seen

 When bioluminescent dinoflagellates such as this one (*Pyrodinium bahamense*) are present in high concentrations, they can literally light up the sea. They respond to pressure changes caused by movement in the water. The word "dinoflagellate" comes from the Greek *dinos* ("whirling") and the Latin *flagellum* ("little whip").

far from any moving objects. Instead, attention turned to bacteria. In 1985, a research vessel sampled water from a milky sea and found that it contained high concentrations of bacteria—a bioluminescent species called *Vibrio harveyi*. Twenty years later, an area that looked like a milky sea was detected in an earlier satellite image and its location off the coast of Somalia in the Indian Ocean correlated exactly with the position of a ship that had reported milky sea conditions at that time. The size of this glowing patch was calculated to be almost 6,000 square miles (more than 15,000 sq km), so the numbers of these single-celled organisms are indeed unimaginably huge.

 Under certain conditions—such as the presence of high levels of nutrients—populations of toxic red algae can increase rapidly, coloring the sea and threatening marine life.

RED TIDE

Other, non-bioluminescent, members of the dinoflagellates can also put on a visual show in the ocean, but with less pleasant consequences. Some species contain brown or red pigment, and when their populations explode, as they do from time to time, they can literally turn the water red over a vast area, a phenomenon that can easily be detected in satellite images. It may look pretty, but it can have serious consequences.

A red tide is just one example of a "harmful algal bloom" or HAB, that can be caused by algae of many different colors, and they are harmful because some species of tiny phytoplankton contain natural neurotoxins, compounds that affect the nervous system if ingested. Under normal circumstances these chemicals don't pose a problem, but the high concentrations that are found in an algal bloom create a hazard for marine life, killing fish, seabirds, and marine mammals. The toxins also accumulate in shellfish such as mussels and oysters, and eating these can cause fatal poisoning.

The causes of HABs, which can also occur in lakes and rivers, are not fully understood. While some appear to be part of a natural cycle, others are undoubtedly linked to human activity, and especially to the pollution of waterways by the runoff of chemical fertilizers from agriculture, which provide the algae with nutrients. Increasing ocean temperatures may also be a factor.

DEAD ZONES

Additional nutrients sound like a good thing, but they have long-term consequences for the oceans. This oversupply of nutrients is known as "eutrophication," and by fueling a huge growth in algae it leads to oxygen depletion in the water. The algae, which produce oxygen in the daytime while they are carrying out photosynthesis, absorb oxygen at night, and they can reduce the oxygen levels in areas of the ocean so greatly that it threatens the lives of other forms of marine life. Even where the level of oxygen is only mildly depleted, this can affect the reproductive capabilities of fish and other creatures. To make matters worse, when the algae die they drift to the sea bed where they are broken down by bacteria in a process that uses up the remaining oxygen, leading to the formation of "dead zones" where nothing can live. As the bodies of creatures killed by the lack of oxygen (such as bottom-dwelling crustaceans and fish that are overcome before they can find oxygenated water) decay, the problem worsens. Since the 1970s, when they were first noticed by scientists, these dead zones have become a focus of research. A United Nations report in 2004 identified 146 such zones, and the number has now almost tripled. The dead zones range in size from a few square miles to the largest known example, to the west of the Mississippi delta in the Gulf of Mexico, which covers more than 8,500 square miles (22,000 sq km).

THE SARGASSO SEA

We have already seen how the central areas of the world's major ocean gyres accumulate floating material, which sadly includes huge quantities of plastic garbage. The coastless Sargasso Sea forms the central portion of the North Atlantic, to the east of Bermuda. This enormous region of clear, slow-moving ocean—an area of some 1.4 million square miles (3.5 million sq km)—is bounded by the fast-flowing Gulf Stream, North Atlantic Drift, and Canary Current, which circulate clockwise around it. Like all the major gyres, it has its share of garbage, but it takes its name from the huge quantities of sargassum seaweed, a form of alga, that float on its surface.

What particularly qualifies the Sargasso Sea as a natural wonder is its intimate connection with a seemingly ordinary, but quite remarkable fish: the eel. Studies showed that the tiny leaf-like larvae found in the Sargasso Sea were not, as previously thought, a distinct species, but grew ultimately into two species of eel, with one found in the waterways of Europe and the other in North America. In neither of these places are larvae are ever found. Although research into the migration of the eels is still incomplete, it is assumed that the mature adults, when they are between 10 and 14 years old, travel as far as 4,000 miles (6,400km) to their spawning grounds in the southern Sargasso Sea, from which they never return. When they are less than half an inch (10mm) long, the tiny larvae make their way into the Gulf Stream and head north, growing as they go. The American eels then turn left while the European eels catch the North Atlantic Drift toward their ancestral home.

Frozen Waters

We have already discovered the wonders of snow and other forms
of frozen precipitation, and now we are going to look at the way
in which these have accumulated in unimaginable quantities in
parts of the globe where the temperature remains below freezing
all year round, and at the strange and beautiful properties of
water when it freezes on the surfaces of oceans, lakes, and rivers.

Ten percent of Earth's surface is perma-
nently covered in ice, some six million
square miles (15.5 million sq km) of it,
comprising almost 70 percent of the world's
fresh water. Not surprisingly, the vast majority of this is to be found in the polar regions,
in the world's great ice sheets, but there are
also great floating shelves of ice, areas of
seasonal sea ice, and glaciers that form in
the world's mountain regions.

ICE SHEETS

An ice sheet is defined as an area of ice-
covered land with an extent of more than
20,000 square miles (50,000 sq km), and
there are only two of these—Greenland in
the northern hemisphere and Antarctica in
the south, both of which exceed the defini-
tion by a long way. The Greenland ice sheet
covers virtually the whole of that island and
has an area of 650,000 square miles (1.7
million sq km), while the Antarctic ice sheet
is more than eight times larger, with an area
of more than five million square miles (13
million sq km). That's almost one and a
half times the size of the United States, and
almost twice the size of Australia.

The volume of ice composing the Antarctic ice sheet is estimated to be more than

7 million cubic miles (30 million cu km),
and it represents the accumulation of some
20 million years of snowfall that has been
compacted under its own weight to form a
dome of solid ice burying the continent's
plateaus and mountain ranges. The average thickness of the ice is about 7,000 feet
(2 km), and in places the ice sheet is twice
that thick. The density within a glacier
changes with depth. The light snowflakes
that fall on the surface become compacted
by further snowfalls and form granular snow,
which contains air that can move around
between the crystals. As it compresses further it becomes *firn*, a type of snow with
smaller crystals and trapped pockets of air.
Finally it turns into ice that becomes ever
denser as more weight is piled on top of it,
the air is compressed into tiny high-pressure
bubbles, and the ice takes on a blue color.

One of the feared consequences of
global warming is the rise in sea level—an
estimated 200 feet (60m)—that would occur
if the ice cap were to melt, releasing huge
water reserves, but with average temperatures around −35°F (−37°C), this is a long
way off.

GLACIERS

In the same way as ice sheets, glaciers are formed by the accumulation and compaction of snow over long periods of time, but glaciers are on the move—huge rivers of ice that flow slowly downhill or are squeezed forward by the pressure above them. In fact, the polar ice sheets actually are glaciers, in that the ice is slowly making its way outward from the thickest areas toward the edges (and we shall see what happens when the ice reaches the sea). Compacted snowfall at high elevations also forms alpine glaciers that flow down mountain valleys, shaping the landscape with the enormous forces that they exert. Although the majority of alpine glaciers and there are literally thousands of them—are to be found in the polar regions, they exist in many of the higher mountain areas of the world. The Rockies, the Himalayas, and the European Alps are obvious examples, but even Mount Kilimanjaro in Kenya has its glaciers.

The gigantic Bering Glacier "flows" into Vitus Lake at the southeastern end of the Alaskan coast. In the last 100 years the tip of the iceberg has retreated some 7 miles (12km).

The longest and largest glacier in North America—118 miles (190km) long and 2,300 feet (700m) thick—is the Bering Glacier in south central Alaska. Flowing from the Wrangell and St. Elias mountain ranges to Vitus Lake, near the Gulf of Alaska, the glacier is right on top of a subduction zone where the Pacific Plate moves beneath the North American Plate. It is thought that the glacier may actually be stabilizing the fault and that as the glacier melts (it is losing seven cubic miles [30 cu km] of water every year) the reduction in weight may lead to increased seismic activity.

The glacier was God's great plow set at work ages ago to grind, furrow, and knead over, as it were, the surface of the earth.

Louis Agassiz

Comprising more than 700 cubic miles (3,000 cu km) of ice, Europe's largest glacier by volume is the Vatnajökull in Iceland. Several volcanoes are buried beneath its massive weight, and Iceland boasts a number of *tuyas*—volcanoes that have flat tops because they erupted beneath glaciers that have since retreated.

ICE SHELVES AND ICEBERGS

Some glaciers flow into lakes or melt to form rivers, but ice-sheet glaciers in the polar regions—Greenland, Canada, and Antarctica—feed into broad floating ice shelves that can be as much as 3,000 feet (1km) thick. Much of the Antarctic coastline consists of ice shelves, with a combined area of about 580,000 square miles (1.5 million sq km). As a glacier adds to a shelf at the line where the land meets the sea (known as the "grounding line"), the shelf extends out into the ocean, and periodically parts of the seaward edge of the shelf will break away, "calving" icebergs. Generally the two processes stay in balance and an ice shelf can remain stable for thousands of years, especially if it is protected from storms and currents by the coastline or by extensive sea ice which, although it is much thinner, prevents waves from eroding the shelf. Sometimes, however, huge sections of an ice shelf can break off or a shelf may collapse completely.

The Ross Ice Shelf is Antarctica's largest ice shelf, 600 miles (950km) across and extending 500 miles (800km) from the coast. In 2000 it calved the world's largest iceberg, a piece of the shelf with an area of some 4,200 square miles (11,000 sq km). This subsequently broke into smaller sections that floated away from Antarctica. Some of them have not yet melted.

Climatologists watch the ice shelves with interest, as their behavior is affected by sea temperature, winter sea-ice extent, and other factors that may be linked to global climate change. There is concern that the collapse of an ice shelf allows the glacier behind it to accelerate, pushing more fresh water into the ocean and raising sea levels.

TIDEWATER GLACIERS

Glaciers that terminate at the coast are known as "tidewater glaciers," and they may calve across a broad front or push a narrow tongue of ice out into the ocean. In either case, they calve relatively small icebergs all the time, sometimes quite dramatically. Icebergs calved from glaciers that reach the coast well above sea level crash down into the sea, while some glaciers, such as the Hubbard Glacier in Alaska, can calve from an edge that is below the water, causing icebergs to burst up through the surface. The impact of a newly calved iceberg can cause a small tsunami, as occurred in Greenland in 1995 when an iceberg-generated wave wrecked boats in a nearby inlet.

SEA ICE

While ice sheets, glaciers, ice shelves, and icebergs are all formed on land and are made of fresh water, sea ice is frozen sea-water, although the salt "distills" out into concentrated brine droplets, leaving salt-free ice. Every winter, the surface of the sea in the polar regions freezes to a greater or lesser degree, extending out into the ocean from the land and from the ice sheets, and every summer it recedes. Like the behavior of the ice sheets, the maximum and minimum extents of the sea ice are of interest to scientists studying the global climate. Whereas the ocean surface absorbs most of the energy that reaches it from the Sun, sea ice absorbs only a small proportion, reflecting the rest. There are concerns that if average global temperatures rise and sea ice is reduced, the warming could accelerate.

The freezing point of fresh water is 32°F (0°C), but seawater freezes at a slightly lower temperature that depends on the salinity. The freezing process goes through several stages. As the water temperature drops to the freezing

Clear freshwater ice coats the surface of Lake Lónið in Iceland. In the background the Gígjökull glacier extends from the Eyjafjallajokull ice sheet.

point, tiny needle-shaped crystals of ice called "frazil" begin to form. Floating in the surface layer, these crystals thicken the water to a slushy consistency and if the sea is smooth they begin to bond together into pieces of thin sheet ice. This gives the sea a smooth appearance, and it is known as "grease ice." The ice then freezes into larger sheets of thin clear ice called "nilas," which gradually thicken and become white in color. Sheets of nilas slide over each other (this is called "rafting") and eventually freeze together to form a thick sheet of "congelation" ice with a smooth lower surface on which ice continues to further accumulate.

In rougher sea conditions, the frazil ice crystals bond together in small circular disks and form "pancake ice." As they knock against each other, these disks develop raised edges. Eventually they freeze together to form sheets, and in rough water these will raft and freeze together in a jumbled form, piling on top of each other and creating a ridged surface. Unlike congelation ice, pancake ice sheets have an uneven lower surface.

Sea ice that forms along coasts and along the edges of ice sheets and adheres to them is called "fast ice," whereas free-floating sea ice is called "drift ice." Both kinds can accumulate to form large areas of pack ice, which is important to species such as the polar bear that need pack ice in order to hunt.

Marine mammals such as beluga whales and narwhals can become trapped in small areas of open water within the pack ice and find themselves too far from the open ocean.

FRESH-WATER ICE CIRCLES

We started our voyage of discovery by gazing out at the seemingly infinite number of spinning galaxies that populate the universe, at the planets that orbit the Sun, and at the Moon that spins around our rotating Earth. We complete our journey by contemplating circular motion on a much smaller scale in the unlikely phenomenon of the ice circle.

Large, rotating, perfectly round disks of thin ice within a circular frame of stationary ice have been observed on rivers in various parts of the world, from Canada and the U.S. to Scandinavia, Russia, and the UK. Theories as to why they occur vary, but the most likely cause is the formation of frazil ice in spinning eddies. As the gently turning pancake of ice grows, usually at a bend in a river where the current swirls slightly, it eventually meets a border of fixed sheet ice that is forming around it. The motion of the central disk prevents the two from freezing to each other and the grinding of one against the other maintains an ice-free, perfectly circular line between them. These weird forms range from a few feet to several yards in diameter, and can take several minutes to complete a revolution.

Under very specific circumstances, when there has been a sudden drop in temperature, smaller free-floating disks with a diameter of a foot (30cm) or so can also appear in open water on lakes and slow-moving rivers.

In the waters off the Antarctic ice shelf, disks of pancake ice cover the surface. They will gradually coalesce to form a large, continuous sheet.

Round, like a circle in a spiral,

Like a wheel within a wheel,

Never ending or beginning

On an ever-spinning reel . . .

from "The Windmills of Your Mind" by Alan and Marilyn Bergman

References

GENERAL REFERENCES

Dunlop, Storm, *The Weather Identification Handbook: The Ultimate Guide for Weather Watchers*, The Lyons Press, Guildford, CT, 2003

Greenberg, Gary, *A Grain of Sand: Nature's Secret Wonder*, Voyageur Press, Minneapolis, MN, 2008

Grenci, Lee M. and Nese, Jon M., *A World of Weather: Fundamentals of Meteorology*. Kendall/Hunt Publishing Company, Dubuque, IA, 2006

Heidorn, Keith C., *And Now... The Weather*, Fifth House, Boston, MA, 2005

Inwards, Richard, *Weather Lore*, BiblioBazaar, Charleston, SC, 2009

Williams, Jack, *The Weather Book*, Vintage Books, London, 1997

Williams, Jack, *The AMS Weather Book: The Ultimate Guide to America's Weather*, University Of Chicago Press, 2009.

Websites:

The Weather Doctor
http://www.islandnet.com/~see/weather/doctor.htm

The Online Meteorology Guide
http://ww2010.atmos.uiuc.edu/%28Gh%29/guides/mtr/home.rxml

Space Weather
http://www.spaceweather.com

Wikipedia
http://www.wikipedia.org (An invaluable resource that provides a solid platform for further research.)

CHAPTER ONE: CELESTIAL PHENOMENA

Night Sky Nation
http://www.nightskynation.com/objects/phenomena

HM Nautical Almanac Office Eclipses Online Portal
http://www.eclipse.org.uk/eclbin/query_eo.cgi

Mr Eclipse
http://www.mreclipse.com/MrEclipse.html

NASA Eclipse Website
http://eclipse.gsfc.nasa.gov/eclipse.html

CHAPTER TWO: OPTICAL PHENOMENA

Greenler, Robert, *Rainbows, Halos and Glories*, Cambridge University Press, 1980

Lynch, David K. and Livingston, William, *Color and Light in Nature*, 2nd edition, Cambridge University Press, 2001

Meinel, Aden and Meinel, Marjorie, *Sunsets, Twilights and Evening Skies*, Cambridge University Press, 1983

Minnaert, M., *The Nature of Light and Colour in the Open Air*, Dover Publications, Mineola, NY, 1948

CHAPTER THREE: ATMOSPHERIC PHENOMENA

Blanchard, Duncan C., *From Raindrops to Volcanoes: Adventures with Sea Surface Meteorology*, Dover Publications, Mineola, NY, 2004

Blanchard, Duncan C., *The Snowflake Man: A Biography of Wilson A. Bentley*, McDonald & Woodward, Blacksburg, VA, 1998

Day, John A., *The Book of Clouds*, Sterling, New York, 2005

Deblieu, Jan, *Wind: How the Flow of Air has Shaped Life, Myth, and the Land*, Houghton Mifflin, Geneva, IL, 1998

Hamblyn, Richard, *The Invention of Clouds: How an Amateur Meteorologist Forged the Language of the Skies*, Farrar Straus & Giroux, New York, NY, 2001

Libbrecht, Kenneth, *Ken Libbrecht's Field Guide to Snowflakes*, Voyageur Press, Minneapolis, MN, 2006

Pretor-Pinney, Gavin, *The Cloudspotter's Guide*, Perigee Trade, New York, 2007

Websites:

Atmospheric Optics Concepts
http://hyperphysics.phy-astr.gsu.edu/hbase/atmos/atmoscon.html

Atmospheric Optics
http://www.atoptics.co.uk

The Cloud Appreciation Society
http://cloudappreciationsociety.org

Ice Ribbons and Frost Flowers
http://my.ilstu.edu/~jrcarter/ice/index-2005.htm

Snow Crystals.com
http://www.its.caltech.edu/~atomic/snowcrystals

CHAPTER FOUR: ELECTRICAL PHENOMENA

Uman, Martin, *Lightning*, Dover Publications,
Mineola, NY, 1984

CHAPTER FIVE: GEOLOGICAL PHENOMENA

National Park Service: Yellowstone
http://www.nps.gov/yell/naturescience/
geothermal.htm

Unavco: Glacial Rebound
http://www.unavco.org/research_science/science_highlights/
glacial_rebound/glacial_rebound.html

University of Lethbridge
http://www.uleth.ca/vft/milkriver/hoodoos.html

Science Clarified
http://www.scienceclarified.com/landforms

Journey Into Amazing Caves
http://www.amazingcaves.com/learn_visit.html

UNESCO: Bradyseism
http://whc.unesco.org/en/tentativelists/2030

ThinkQuest: Landslides
http://library.thinkquest.org/C003603/english/landslides

CHAPTER SIX: AQUATIC PHENOMENA

Dai, Aiguo and Trenberth, Kevin E., "Continents: Latitudinal
and Seasonal Variations," in the Journal of Hydrometrology,
Volume 3, December 2002

Websites:

Fisheries and Oceans Canada: Tidal Phenomena
http://www.charts.gc.ca/twl-mne/general-generales/
phenomen-eng.asp

Meteorological Service of Canada: El Niño
http://www.msc.ec.gc.ca/education/elnino/index_e.cfm

University of Victoria, Canada: Dead Zones
http://communications.uvic.ca/uvicinfo/announcement.
php?id=397

National Oceanographic and Atmospheric Administration:
Dead Zones
http://www.noaanews.noaa.gov/stories2009/pdfs/new%20
fact%20sheet%20dead%20zones_final.pdf

University of California Santa Barbara: Dead Zones
http://www.lifesci.ucsb.edu/~biolum/organism/milkysea.html

The National Snow and Ice Data Center, University of
Colorado at Boulder
http://nsidc.org

Index

Image Credits

Back cover:
    top   © Mike Norton | Dreamstime.com
Front cover:
    © NASA Johnson Space Center – Earth Sciences and
    Image Analysis
    (Outside Flap) © Delamofoto | Dreamstime.com
Flap:
    top-left © Romko | Dreamstime.com
    top-right © Gordon Garradd | Science Photo Library
    bottom-left © Mircea Madau
    bottom-center © Wangleon | Dreamstime.com
    bottom-right © Larry Landolfi | Science Photo Library
9   © Win Initiative | Getty
12   © NASA and the European Space Agency
15   © NASA/JPL-Caltech/S. Stolovy (SSC/Caltech)
16   © European Space Agency | Science Photo Library
20   © Getty Images
22   © John Chumack | Science Photo Library
23   © Navicore | Creative Commons
25   © Sanchezn | Creative Commons
26   © Bowie15 | Dreamstime
29   © Emiliano Ricci | Creative Commons
30   © Eckhard Slawick | Science Photo Library
33   © First Light | Alamy
34   © Wangleon | Dreamstime.com
36   © Russell Bernice | Creative Commons
38   © NASA
42   © Sad444 | Dreamstime.com
44   © Jeremy Hohengarten | iStockphoto
47   © Creative Commons | Mila Zinkova
49   © Stephen & Donna O'Meara | Science Photo Library
51   © David R. Tribble | Creative Commons
53   © Mila Zinkova | Creative Commons
55   © Constantine | Creative Commons
56   © Gordon Garradd | Science Photo Library
61   © Vicnt | Dreamstime.com
65   © NOAA
66   © imagebroker | Alamy
69   © Mila Zinkova | Creative Commons
71   © Frank Zullo | Science Photo Library
72   © Roman Krochuk | Dreamstime.com
75   © 36clicks | Dreamstime.com
77   © Joss | Dreamstime.com
80   © Jeff Schmaltz, MODIS Rapid Response Team,
    NASA/GSFC
85   © NOAA George E. Marsh Album
89   © Stephen Frink
90   © Dr. Joseph Golden, NOAA
92   © Axel Rouvin
95   © Smari | Getty
96-7   All Public Domain except:
    Cumulus congestus: © Simonhs | Dreamstime.com
    Cumulus humilis: © Selectphoto | Dreamstime.com

100   © John2165 | Creative Commons
104   © Andrew Penner | iStockphoto
107   © Tenetsi | Dreamstime.com
109   © Jiri Kadlec
113   © Keith Heidorn
115   © Michael | Creative Commons
116   © Sipuli | Dreamstime.com
119   © Taro Taylor | Creative Commons
121   © Keith Heidorn
123   © Victor Melniciuc | iStockphoto
127   © Popperfoto
129   top © Public Domain
129   bottom © Slomoz | Creative Commons
132   © Fotolotti | Dreamstime.com
137   © Clint Spencer | iStockphoto
139   © Mircea Madau
143   © Wizzard | Dreamstime.com
145   © Public Domain
147   © Arctic-Images | Corbis
149   © InterNetwork Media | Getty
150   © Daniel L. Osborne, University of Alaska | Detlev Van
    Ravenswaay | Science Photo Library
154   © Mike Hollingshea | Science Faction | Corbis
156   © Ji Elle | Creative Commons
159   © SSPL via Getty Images
162   © Argironeta | Dreamstime.com
164   © Public Domain
166   © Marco Soave
167   © Game McGimsey | USGS
169   © Library of Congress
170   © javarman3 | iStockphoto
173   © Bcbounders | Dreamstime.com
176   © George Burba | Dreamstime.com
178   © Creative Commons | H. Raab
180   © Alan Kearney | Getty Images
182   © Creative Commons | Håvard Berland
186   © Tomper | Dreamstime.com
188   © Dylan Kereluk | Creative Commons
192   © Public Domain
197   © Latin Content | Getty Images
198   © Norman Kuring | NASA Earth Observatory
200   © Delamofoto | Dreamstime.com
204   © Dea | G. Dagli Orti
207   © Jocelyn Augustino | FEMA
208   © Terry Hazen, Visuals Unlimited | Science Photo Library
210   © Carleton Ray | Science Photo Library
213   © NASA
215   © Andreas Tille | Creative Commons
216   © Michael Clutson | Science Photo Library